ACS SYMPOSIUM SERIES **448**

Microemulsions and Emulsions in Foods

Magda El-Nokaly, EDITOR
The Procter and Gamble Company

Donald Cornell, EDITOR
U.S. Department of Agriculture

Developed from a symposium sponsored
by the Division of Agricultural and Food Chemistry
at the 199th National Meeting
of the American Chemical Society,
Boston, Massachusetts,
April 22–27, 1990

American Chemical Society, Washington, DC 1991

Library of Congress Cataloging-in-Publication Data

Microemulsions and emulsions in foods / Madga El-Nokaly, editor;
Donald Cornell, editor.
p.   cm.—(ACS Symposium Series; 448).

"Developed from a symposium sponsored by the Division of
Agricultural and Food Chemistry at the 199th Meeting of the American
Chemical Society, Boston, Massachusetts, April 22–27, 1990."

Includes bibliographical references and index.

ISBN 0–8412–1896–X
1. Emulsions—Congresses.   2. Food—Congresses.   3. El-Nokaly,
Magda A., 1945–   . II. Cornell, Donald. 1931–   . III. American
Chemical Society. Division of Agricultural and Food Chemistry.
IV. American Chemical Society. Meeting (199th: 1990: Boston,
Mass.). V. Series.

TP156.E6M513   1991
664—dc20   90–22719
                                     CIP

The paper used in this publication meets the minimum requirements of American National
Standard for Information Sciences—Permanence of Paper for Printed Library Materials, ANSI
Z39.48–1984.   ∞

TP 156
E6
M513
1991
CHEM

ACS Symposium Series

M. Joan Comstock, *Series Editor*

1991 ACS Books Advisory Board

Foreword

THE ACS SYMPOSIUM SERIES was founded in 1974 to provide a medium for publishing symposia quickly in book form. The format of the Series parallels that of the continuing ADVANCES IN CHEMISTRY SERIES except that, in order to save time, the papers are not typeset, but are reproduced as they are submitted by the authors in camera-ready form. Papers are reviewed under the supervision of the editors with the assistance of the Advisory Board and are selected to maintain the integrity of the symposia. Both reviews and reports of research are acceptable, because symposia may embrace both types of presentation. However, verbatim reproductions of previously published papers are not accepted.

Contents

Preface

THE FIELD OF FOOD EMULSIONS HAS ALWAYS GENERATED great interest. However, little has been done with microemulsions applicable to the complex world of foods. Microemulsions have been the subject of much fundamental research that focuses on noningestible systems. For example, applications to nonfood uses such as tertiary oil recovery, fuel, cosmetics, and household products has received considerable attention.

The purpose of this book is to bring together the two related disciplines of emulsions and microemulsions in an attempt to foster new developments and new directions for fundamental research. Although the book is intended primarily for technologists in the food industry and researchers in academe, much of the knowledge contained herein may be directly applicable to cosmetics, pharmaceutical areas such as health care or drug delivery, and stabilization of lotions and creams.

The introduction, written by Paul Becher, analyzes the book's chapters, and the overview by Stig Friberg and Ibrahim Kayali examines the physical differences between microemulsions and emulsions.

The first section of the book covers microemulsions in foods. The scientific literature contains little information on microemulsion systems directly applicable to foods. Kare Larsson's monoglyceride/water/oil system was the first practical system to be published. Other chapters describe synthesis in microemulsion media and preparation of microemulsions and phase diagrams.

The final section of the book looks at emulsions in foods. The phase behavior of sucrose esters is discussed, and the emulsion stabilization properties of newly synthesized polypeptides are reported. The largest share of the book is on emulsions stabilized by proteins: protein–emulsifier and protein–protein interactions and their effects on interfacial films. Stabilization of emulsions by polysaccharide is also represented. Novel nondestructive methods for measuring emulsion stability, with broad applications in emulsion technology, are described.

Overall, the book covers many of the more innovative approaches to emulsion stability in foods, presents the results of some initial investigations into food microemulsions, and compares the two types of systems. It is hoped that the book will inspire others to continue the investigation of food microemulsions.

Acknowledgments

We would like to thank the ACS Agricultural and Food Chemistry Division program chairman, T. E. Acree, and treasurer, C. J. Mussinan, for their support. We are also grateful for the partial financial support provided by Procter and Gamble, with special thanks to Ted Logan of the Ph.D. recruiting office and to Joseph McGrady of the Food and Beverage Technology Division.

We would like to thank each of the contributing authors for their cooperation, without which there would not have been a book. Many thanks to the editorial staff of the ACS Books Department, especially to Cheryl Shanks and Barbara Tansill. Last but not least, we would like to acknowledge with thanks Carla Cobb and Pat Greene for their secretarial support.

MAGDA EL-NOKALY
The Procter and Gamble Company
Cincinnati, OH 45239–8707

DONALD G. CORNELL
Eastern Regional Research Center
U.S. Department of Agriculture
Philadelphia, PA 19118

August 28, 1990

Chapter 1

Food Emulsions

An Introduction

Paul Becher

Paul Becher Associates Ltd., P.O. Box 7335, Wilmington, DE 19803

In this introduction the long history of food emulsions is first con-
trasted with the comparatively short time over which they have
been subject to scientific study. The contrasting degree of interest
in macro– and microemulsions in the food industry is discussed, and
the various ways in which such emulsions may be studied is considered. Finally, the various aspects of this symposium are considered,
as well as the role which some of the contributions may play in the
progress of this discipline.

It is an interesting fact that the first food emulsion we encounter during
our lives is also probably the oldest — namely, milk. Mammalian life is first
encountered several billion years after the Big Bang, probably of the order of
one hundred million years ago, and, with mammals, comes milk. In Table I I
have presented a short (and, obviously, abbreviated) history of food emulsions.
However abbreviated, it is nonetheless useful as indicating that the subject of
this symposium does have a long history.

However, in spite of this long history, the scientific study of emulsions in
general, and food emulsions in particular, is a fairly recent phenomenon. The
word *emulsion* dates from the first years of the 17th century (although not
precisely with the current connotation), but the scientific investigation is little
more than a century old; indeed, the classification of emulsions as oil-in-water
or water-in-oil dates from a 1910 paper by Wolfgang Ostwald [1].

It would be even more correct to say that the true theoretical under-pinnings
of emulsion theory could not exist until after the post-WWII development of
the theory of the electric double layer, the DLVO theory [2].

Even more recent, of course, is the concept of the microemulsion, dating from
Schulman and Montagne in 1961 [3]. Although the title of this symposium is,

0097–6156/91/0448–0001$06.00/0
© 1991 American Chemical Society

2   MICROEMULSIONS AND EMULSIONS IN FOODS

Table I: A Short History of Food Emulsions

FOOD	DATE OF INTRODUCTION
Mammalian Milk	c. 2.4×10^7 BC
Milk from Domestic Animals/Butter/Cheese	c. 8500 BC
Sauces	15th–16th Century
Ice Cream	c. 1740
Mayonnaise	c. 1845
Margarine	1869

in fact, *Microemulsions and Emulsions in Foods*, and, although many of the papers included in the symposium are on various aspects of microemulsions, I suspect that we will find that this is a new field, with much to be said for the future, but little we can say about the past.

In fact, in preparation for writing this paper, I did a quick and dirty computer search based on the simple research strategy

food? and microemulsion?

The yield from this search was a mere four references (two to patents), none older than 1987. On the other hand, the search strategy

food? and emulsion?

yielded no less than 475 hits.

I should point out, in addition, that a recent review of applications of microemulsions contained but a single reference to foods [4].

In fact, a number of recent books on food emulsions, edited by, respectively, Friberg (1976) [5], Dickinson (1987) [6], and Dickinson and Stainsby (1988) [7], contain no instances of food microemulsions (except for one minor reference in [6]).

This lack of interest may possibly be ascribed to a number of related factors. First, the high levels of emulsifying agents normally encountered in microemulsions quite simply serves as an economic barrier. Second, this same high level of emulsifier might well raise legal problems in securing approval from the FDA. Third, of course, there is the simple fact that it is apparently quite difficult to make microemulsions of the fats and oils used in foods.

Why, then, a symposium on *Microemulsions* and Emulsions in Foods? As far as *macro*emulsions are concerned, the considerable activity noted above is reason enough – we simply want to keep posted. On the other hand, the inclusion of micro emulsions is a matter of looking to the future. Is there, in fact, a role for microemulsions in the food industry, and, if so, what do food technicians have to know about this topic?

The editors have chosen, logically enough, to order the papers in this symposium as follows:

- Introduction (this paper)

- Overview

- Microemulsions in Foods

- Emulsions in Foods

For the purposes of this introduction, however, I would like to consider the various contributions in a slightly different way:

• Micro– and macroemulsions, without specific reference to food, serving as background.

• Food emulsions, principally (but not exclusively) macroemulsions.

• Food emulsifiers, their properties or production.

Following the lucid overview of Friberg and Kayali (which defines further the distinctions between micro– and macroemulsions, and illuminates the difficulties involved in the application of microemulsions to foods alluded to above), in the first category, we may single out the papers of Larsson; El-Nokaly; Osborne, Pesheck, and Chipman; and Biais, Bothorel, Clin, and Lalanne. In particular the last of these connects the concept of the influence of the bending energy with stability in microemulsions. This was also touched upon by Friberg and Kayali, which with the papers in this group, provide an introduction (albeit a brief one) to the problems associated with microemulsions and their formulation.

There is more meat (if one may be permitted a gustatory pun) in the second category. The interaction between protein and emulsifiers (a significant effect in many food emulsions) is discussed from various points of view by Dickinson; Martinez Mendoza and Sherman; Westerbeek and Prins; Li-Chan and Nakai; and by Makino and Moriyama. It is interesting to note that in various of these papers the investigators have favored a different research technique, in all cases with illuminating results. Two other papers, one by Goetz and El-Aaser, the other by Robins, are also concerned with investigative techniques; the former uses electroacoustic measurements to determine ζ–potentials, while the latter employs an ultrasonic monitor to follow creaming.

In the final category, there useful papers on the synthesis of monoglycerides in microemulsion (Mazur, Hiler, and El-Nokaly); sucrose esters, describing their L_2 phases (Herrington, Midmore, and Sahi); liposarcosine polymeric surfactants, which form microemulsions with the formation of liquid crystals (Gallot); protein–dextran hybrids (Kato and Kobayashi); and stabilizers for milk–derived vesicles (Whitburn and Dunne).

As my own contribution to this symposium, I would like to include a few words on the subject of research in foods.

In view of the lengthy history of food emulsions detailed in Table 1, the relatively recent attention to the theory of food emulsions is illustrated by the classic 1946 paper of Corran [8] on mayonnaise. Corran investigated the effect of the egg-yolk composition (lecithin-cholesterol ratio), phase volume, emulsifying effect of the mustard (effect of fineness), method of mixing, water hardness, and the viscosity of the finished product. Although Corran's methods were primitive by today's standards, the completeness of the study can stand as a paradigm for the investigation of food emulsions. This paper, or the summary in Becher [9], may still be referred to usefully.

Although the composition of mayonnaise is strictly regulated by law, it should be pointed out that true mayonnaise aficionados claim to be able to distinguish between various commercial brands. Verification of this, and, if true, an explanation, would represent a proper homage to Corran, as well as a significant piece of food research.

As another significant bit of food research (even if tongue–in–cheek), would be the famous exchange of papers regarding the application of DLVO theory to Béarnaise sauce [12].

The study of food emulsions may be considered to have followed three paths:

- Physico–chemical studies of the whole emulsion system, e.g., effect of composition on stability, rheology, color, taste, etc.

- Physico–chemical studies of model systems, e.g., surfactant interactions with other components, etc.

- Formulation of finished emulsion products, including consumer test panel evaluation, packaging studies, etc.

On the first such path, the food itself has been the subject studied (as done by Corran), e.g., milk, ice cream, cake emulsions, meat emulsions, etc. On the second path, model systems have been investigated, although these studies have often been limited to a single interaction, e.g., water with lipids, surface-active agents with protein, etc. A number of examples of this approach will be found in the present symposium.

A third path may be considered to exist: I refer to the actual formulation of food systems, wherein, one may hope, the formulator is guided by the lessons derived from the first two paths.

As an example of the first route we may consider milk One may possibly be surprised to see such an ancient system as milk (Table I) listed as those subject to investigation A large part of the studies on milk, of course, have been devoted to homogenization [10]. (I wonder how many of this group remember the days of *un*homogenized milk, when there was a premium in rich coffee cream for the first one up!), with possibly as much attention devoted to the nature of the milk interface [11].

I should emphasize, also, that in food emulsions the relation between the properties of the emulsion and its texture (e.g., mouth feel in such systems as ice cream) is also a major area for investigation.

The second route particularly is exemplified by numerous studies of the behavior of proteins and surfactants at the oil/water interface, of which excellent examples are to be found in the recent literature [6, 7].

As for the third route — formulation — the results of these investigations will be found widely throughout the patent literature, and, of course, on the shelves of your local supermarket.

References

[1] Ostwald, Wo. *Kolloid-Z.* 1910, *6*, 103.

[2] (a) Derjaguin, B. V.; Landau, L. *Acta Physiochem. USSR* 1941, *14*, 633; (b) Verwey, E. J. W.; Overbeek, J. Th. G. *Theory of the Stability of Lyophobic Colloids*; Elsevier: Amsterdam, 1948.

[3] Schulman, J. H.; Montagne, J. B. *Ann. N.Y. Acad. Sci.* 1961, *92*, 366.

[4] Gillberg, G. In *Emulsions and Emulsion Technology*; Lissant, K. J, Ed.; Marcel Dekker, Inc.: New York and Basel, 1984; Part III, pp. 1-43.

[5] Friberg, S., Ed. *Food Emulsions*; Marcel Dekker, Inc.: New York and Basel, 1976.

[6] Dickinson, E. *Food Emulsions and Foams*; Royal Society of Chemistry (Spec. Pub. No. 58): London, 1987.

[7] Dickinson, E.; Stainsby, G. *Advances in Food Emulsions and Foams*; Elsevier Applied Science: London and New York, 1988.

[8] Corran, J. W. In *Emulsion Technology*; Chemical Publishing Co.: Brooklyn, N.Y., 1946; pp. 176-192.

[9] Becher, P. *Emulsions: Theory and Practice*, 2nd ed.; Robert E. Krieger Publishing Co.; Melbourne, FL, 1977 (Reprint of 1966 edition); pp. 344-347.

[10] Walstra, P. In *Encyclopedia of Emulsion Technology*; Becher, P., Ed.; Marcel Dekker, Inc.: New York and Basel, 1983; Vol. 1, Basic Theory, Chap. 2.

[11] Graf, E.; Bauer, H. In *Food Emulsions*: Friberg, S., Ed.; Marcel Dekker, Inc.: New York and Basel, 1976; Chap. 7.

[12] (a) Perram, C. M.; Nicolau, C.; Perram, J. W. *Nature* 1977, *270*, 572-573; (b) Small, D. M.; Bernstein, M. *New England J. Med.* 1979, *300*, 80-802; (c) Walker, J. *Scientific American, 1979*(12), 178-190.

RECEIVED August 16, 1990

Chapter 2

Surfactant Association Structures, Microemulsions, and Emulsions in Foods

An Overview

Stig E. Friberg and Ibrahim Kayali

Chemistry Department, Clarkson University, Potsdam, NY 13699–5810

A great variety of dispersants are employed in food emulsions ranging from hydrotropes to solid particles. However, surfactant association structures, micelles, and lyotropic liquid crystals are of decisive importance for the properties of food emulsions and their response during the absorption of lipids in the stomach. Examples are provided demonstrating the use of hydrotopes and lamellar liquid crystals to form microemulsions with liquid triglycerides. In addition the importance of liquid crystals for emulsion is demonstrated.

Emulsions are well known in food science and several monographs have focused on this theme (1-3). A large variety of stabilizers are used to obtain optimal properties to the emulsions depending on the final product (4). In addition the application of specific stabilizers is related to the natural origin of food products; in fact some emulsions are of biological origin.

Such an emulsion is milk, a 4% fat O/W emulsion is an aqueous phase with a few percent protein and lactose. It is stabilized by a lipid protein membrane, whose structure has been extensively investigated (5-7). The early suggestions of a combination of a phospholipid bilayer and proteins (7) appears to have stood the test of time. However, milk in its native state is not available commercially and the research has focused on changes in the membrane due to processing (8). The kinetic interfacial processes with the competition for surface sites between proteins and lipids have generated a research using different approaches (9,10). Other food dispersions (4) stabilized by proteins are ice cream (caseinate), cake batter (whey protein), and mayonnaise (egg protein). Polysaccharides are another natural protein applied to artificial creams and salad dressings.

0097–6156/91/0448–0007$06.00/0

The proteins act as polymer stabilizers (11) but the ones with a compact conformation precipitate to form small particles. Emulsion stabilization by small particles was early described by Pickering (12) and has been reviewed (4,13). The stabilization is based on the wetting energy increase when the small particles are forced from their equilibrium location at the interface into the dispersed phase during coalescence. The wetting energy is greater the higher the contact angle in the dispersed phase but in practice a value under but close to 90° has been found optimal.

The value of the contact angle is influenced by low molecular weight surfactants and their presence is essential in special applications such as margarine, bakery products and toppings.

However, to contrast emulsions and microemulsions it is necessary to consider cases in which emulsions are stabilized by a low molecular surfactant such as lecithin or a monoglyceride. Microemulsions *per se* have not been treated much in food products, mainly because the standard microemulsion formulation schemes (14-16) are not well suited for triglyceride systems.

The microemulsions and emulsions in both systems are related because surfactant association structures are decisively involved. These relations are not generally well known, a fact that causes serious problems in the formulation efforts. Hence, we considered a review of the phenomena involved and their relations to be of general value.

The treatment includes a short description of surfactant amphiphilic structures, a discussion on microemulsion/emulsion structure/stabilization by surfactants and a summary of liquid crystal/emulsion relations.

Micelles and Lyotropic Liquid Crystals.

A water soluble surfactant adsorbs strongly to an interface toward the air or toward an oil because of its dual structure with a hydrocarbon tail with insignificant interaction with the water (the hydrophobic part) and a polar group with strong interaction with water (the hydrophilic part). This adsorption causes a reduction of the interfacial free energy. An oil soluble surfactant does not adsorb toward the oil/air interface, but does so toward an oil/water interface.

For surfactant concentrations above a certain limit in water (the critical micellization concentration, c.m.c.) the added surfactant forms micelles, Figure 1, and the adsorption to the interface does not increase with surfactant concentration. In an oil, the oil soluble surfactants and water form inverse micelles, Figure 2, in a step-wise process.

These two structures are especially important in systems in which an ionic surfactant and a long chain alcohol are combined, because this system illustrates, with great clarity, the essential difference between the stabilizing system for a microemulsion and an emulsion. Once this difference is distinguished the difficulties with microemulsions in food products are easy to comprehend.

A water/ionic surfactant/alcohol system is shown in Figure 3. The aqueous micellar solution, A, dissolves (solubilizes) some alcohol and the alcohol solution dissolves huge amounts of water into the inverse micelles, B. These two areas are not in equilibrium with each other, but are separated by a third region, a lamellar liquid crystal.

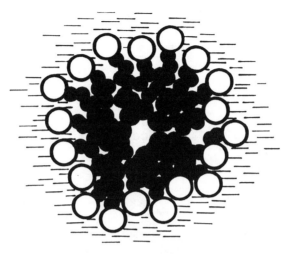

Figure 1. In a normal micelle the surfactant hydrocarbon chains (black) point toward the inner part surrounded by the polar parts (unfilled circles).

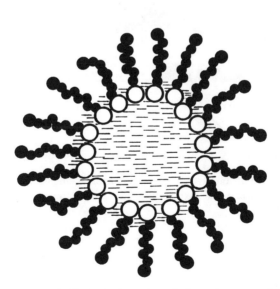

Figure 2. In an inverse micelle the hydrocarbon chains point outward while the polar groups are concentrated in the center.

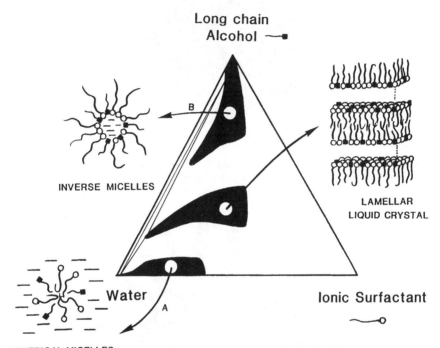

Figure 3. In a combination of water, a surfactant and a long chain alcohol, the areas for solutions of normal and inverse micelles, are separated by a lamellar liquid crystal.

This phase and its equilibria with the aqueous micellar solution (A) and the inverse micellar solution (B) are the essential elements for both microemulsion and emulsion stability. They are discussed in the following sections.

Microemulsions versus Emulsions

The most characteristic difference between an emulsion and a microemulsion is their appearance. An emulsion is turbid while the microemulsion is transparent, Table I.

Table I. Characteristics of Emulsions and Microemulsions

	Emulsion	Microemulsion
Appearance	Turbid	Transparent
Droplet size, μm radius	0.15 - 100	0.0015 - 0.15
Formation	Mechanical or Chemical Energy added	Spontaneous
Thermodynamic Stability	No	Yes (No)

The reason for this difference in appearance is the size of the droplets. For an emulsion the droplets are of dimensions similar or greater than the wave length of light and light is reflected off the droplets. The emulsion, hence, appears turbid because the light cannot penetrate through it. The size of microemulsion droplets is smaller than the wave length of light, and the interaction with light is limited to scattering. A light beam passes through with but little loss; the microemulsion appears transparent.

The microemulsions are thermodynamically stable with few exceptions. They form spontaneously; it is obvious that no method of stirring would bring droplet size to a magnitude of a few tens of Ångströms. Emulsions, on the other hand, are not thermodynamically stable because the interfacial energy is positive and dominant in the total free energy. The essential difference is presented in Figure 4; the surface free energy of the microemulsion has two components stretching (positive contribution) and bending (negative contribution). The two cancel each other and the total surface free energy is extremely small $\approx 10^{-3}$ mN/m. The emulsion droplet is of a size that the bending energy is negligible and the surface free energy is large and positive; a few mN/m.

Recalling the diagram in Figure 3, it is now clear that the microemulsions are related to the micellar solutions A and B. In fact, the W/O microemulsions are obtained by adding a hydrocarbon to the inverse micellar solution B, Figure 5, while the O/W microemulsion region emanates from the aqueous micellar solution A, Figure 3. The essential information is that the microemulsion regions are in equilibrium with the lamellar liquid crystal.

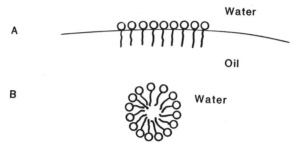

Figure 4. The curvature in an emulsion droplet (A) is extremely small and the bending component of the surface energy is not significant. A change in curvature does not lead to a change in the free energy. In the microemulsion droplet (B), on the other hand, a change in curvature leads to a pronounced change in free energy; e.g., the bending component of the surface free energy is pronounced.

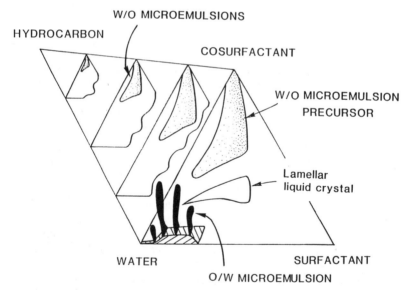

Figure 5. The W/O microemulsion region emanates from the inverse micellar solution and the O/W area from the normal micellar solution. The lamellar liquid crystal separates the two microemulsions.

This phase has to be destabilized to an optimal degree in order to maximize the microemulsion regions. Hence, a cosurfactant is chosen with a limited chain length; approximately C5, to destabilize the lamellar packing of the liquid crystal to form small spherical droplets. For an emulsion, on the other hand, with its huge radius the parallel packing of the surfactant/cosurfactant is optimal. Hence, the cosurfactant should now be of a chain length similar to that of the surfactant - all in accordance with practical experience (16).

Food Systems, Microemulsions

From these considerations it follows automatically that a surfactant/cosurfactant combination, which is optimal for a microemulsion, is of little use in order to stabilize an emulsion. This is a serious disadvantage when double emulsions, W/O/W, are formulated, Figure 6. For a system of this kind a W/O microemulsion emulsified into water would in principle be a very attractive option, because the W/O part, which is the difficult part to stabilize would now be thermodynamically stable.

However, stabilization using surfactant combinations has fundamental difficulties. The surfactant combination for the microemulsion will rapidly exchange with the one for the emulsion, which leads to destablization for both the emulsion and the microemulsion.

The emulsion stabilizers favor a lamellar packing of surfactant at the interface and are, hence, not suitable for the microemulsion; they will reduce the maximum solubilized water in it drastically forming instead an unstable W/O emulsion. In addition the microemulsion combination will destroy the packing of surfactant/cosurfactant around the emulsion droplets; the layer will become more disordered and mobile and O/W emulsion stability is lost also.

This dilemma has been resolved in an elegant manner by Larsson *et. al* (17). They used a surfactant combination to stabilize the W/O microemulsion but avoided the problem of the emulsion part by using a polymer as its stabilizing agent. The polymer being water soluble, is virtually insoluble in the oil part of the microemulsion and its dimensions prevents its inclusion into the W/O droplets. It will, hence, not interfere with the W/O microemulsion stabilization system.

In addition the polymer adsorbs strongly to an interface (18), thereby eliminating the competition about the emulsion interface from the surfactant combination. A very elegant solution to a difficult problem.

As pointed out earlier liquid triglycerides do not lend themselves to microemulsion formulation with the traditional technique, Figure 5. Addition of liquid triglycerides to the inverse micellar solution results in a phase change to a lamellar liquid crystal. Hence, a different strategy must be employed in order to prepare a microemulsion with triglycerides. One solution is to attack the problem from the opposite side. Realizing that the microemulsions are obtained by destabilizing a liquid crystal, Figure 3, it appears reasonable to approach the problem with a liquid crystal as starting point instead. This means forming a liquid crystal containing triglycerides and destabilizing it by addition of a suitable compound.

This compound should be considered as a potent cosurfactant destabilizing the liquid crystalline phase. The common cosurfactants are, however, not useful because of

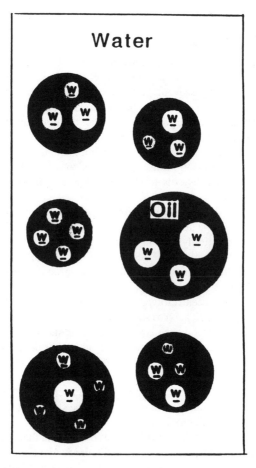

Figure 6. In a W/O/W emulsion, water droplets are dispersed in oil drops, which, in turn, are dispersed in the continuous aqueous phase.

their toxicity. The destabilization was instead obtained by the use of hydrotropes (19). These are compounds, the action of which is the destabilization of liquid crystals as was early demonstrated (20). A large number of them are allowed into food products.

The lamellar liquid crystals used for the experiments used monoglycerides or lecithin as the stabilizer (19). Figs. 7A-C show the isotropic liquid region obtained by addition of a hydrotrope to a lamellar liquid crystal. The solution in the system of water/1-monocaprylin/sodium xylene sulfonate, Figure 7B, is the largest one and it was used to dissolve a triglyceride, trioctanoin.

The results, Figure 8, illustrate the limitations found. The amount of triglyceride dissolved was very limited indeed with a maximum of 13.5% by weight of triglyceride. Other approaches have resulted in similar results. So, for example, does the system water/monocaprylin/tricaprylin (21), Figure 9, shows very little water solubilization into the oil. To reach 15% by weight of water 50% of the stabilizers were needed.

Emulsion Applications

The request that a stabilizer for emulsions should pack in a plane layer leads to a frequent occurrence of lamellar liquid crystals in the phase diagrams of food emulsifiers (22). So has the glyceryl monoesters huge areas of lamellar liquid crystals as has lecithin.

In many cases the lamellar liquid crystal will remain stable also in the presence of both water and oil, and a three-phase emulsion is obtained. The third phase is a lamellar liquid crystal and it serves as a stabilizer for the emulsion with several mechanisms. These have been described in individual publications (23-25), but a systematic comparison of them may be useful.

As Barry (23) has pointed out the liquid crystal will be finely dispersed in the continuous phase thereby increasing the viscosity of it. An increase in viscosity leads to a reduction in the flocculation rate:

$$\frac{dn}{dt} = \frac{-8kT}{3\eta}n^2/2a \;_{2a}\!\!\int^{\infty} e^{v/kT}\frac{dl}{2} = \frac{A}{B}n^2 \qquad [1]$$

in which n is the number of droplets, t time, k Boltzmann's constant, T absolute temperature, η viscosity, a radius of droplet, v interdroplet potential, and l the distance between droplet centers. The time for one half of the droplets to disappear through aggregation is

$$t_{1/2} = B/An_0 \qquad [2]$$

For an unprotected emulsion, one cubic centimeter of an emulsion with an oil-to-water ratio of 1/1, the $t_{1/2}$ is of the order of one second.

The $t_{1/2}$ is proportional to the viscosity; eqs. [1] and [2] and it is obvious that viscosity as such cannot reach values ($6 \cdot 10^5$ P ') to bring reasonable stability to an

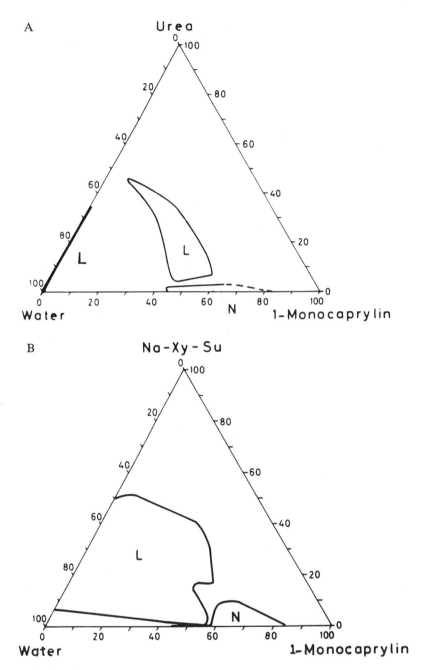

Figure 7. Phase diagrams of aqueous systems of water, surfactant, and hydrotope.
L = isotropic liquid; N = lamellar liquid crystal. (A) Surfactant, 1-Monocaprylin;
Hydrotope, Urea. (B) Surfactant, 1-Monocaprylin; Hydrotope, Sodiuim xylene sulfonate.
(Figure 7 continued on next page)

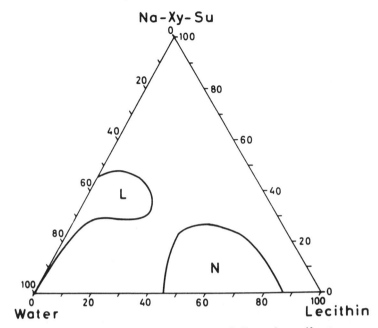

Figure 7C. Surfactant, Lecithin; Hydrotope, Sodium xylene sulfonate.

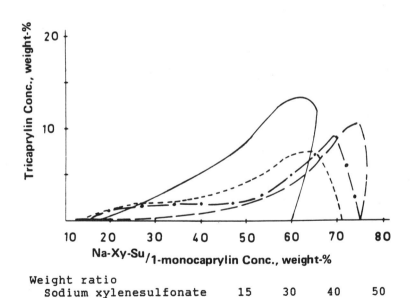

Weight ratio				
Sodium xylenesulfonate	15	30	40	50
Monocaprylin	85	70	60	50

Figure 8. Tricaprylin solubilization in the isotropic solution in Figure 7B. (Reprinted with permission from ref. 19. Copyright 1971 American Oil Chemists' Society.)

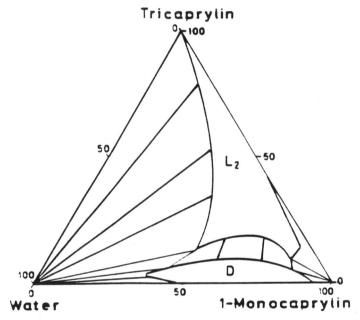

Figure 9. The W/O microemulsion region (L₂) in the
system water, 1-monocaprylin and tricaprylin
solubilizes only small amounts of water. (Reprinted
with permission from Friberg, S. E.; Mandell, L. J. Am.
Oil Chem. Soc. 1970, 47[5], 150. Copyright 1970
American Oil Chemists' Society.)

emulsion. Such an emulsion would have no practical value except as a cream. Instead of pure viscosity the liquid crystal rather should be considered as forming a three-dimensional network in the continuous phase (23). Such a network will be too weak to affect the macrorheological properties much, but serves to provide a yield value against the feeble thermal forces giving the movement of the emulsion droplets.

It is essential to realize that the amount of liquid crystal is not proportional to the amount of emulsifier. Table II shows the amount of liquid crystal for different emulsifier concentrations in the one-to-one oil-and-water emulsion in Figure 10.

Table II. The Amount of Liquid Crystal for Different Emulsifier Concentrations in the 1/1 Oil and Water Emulsion in Figure 10

Emulsifier concentration weight % of total	Liquid crystal weight percent of total
2	0
3	2.3
5	14.6
10	42.9
15	75.4

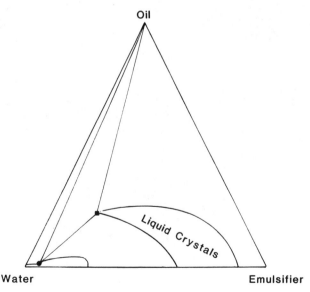

Figure 10. A typical phase diagram of an emulsion system containing a liquid crystal.

With a moderately low emulsifier content in the liquid crystal the amount of stabilizing compound reaches huge amounts even for modest emulsifier concentrations.

Simple geometry gives an analytic expression for the amount of liquid crystal for the case with limited solubility of the surfactant in one phase (water) = s_w weight fraction in the aqueous phase, the surfactant concentration equal to s_{Lc} in the apex of the liquid crystal and an oil/water weight ratio of R for that point (R ≥ 1). The expression for the weight fraction liquid crystal is if s_s is the weight fraction of surfactant and the oil-water weight ratio of the emulsion equals one.

$$m_{LC} = (2s_s - s_w) / \left[\frac{2s_{LC}(1+R)}{2R + s_{LC}(1-R)} - s_w \right]$$ [3]

The lamellar liquid crystal has a typical pattern, when viewed between crossed polarizers, Figure 11. This pattern cannot be observed in detail when the liquid crystal is part of an emulsion, but the radiance of the adsorbed liquid crystal is visible also in that state, Figure 12. Electron microscopy allows the individual layers to be observed, Figure 13.

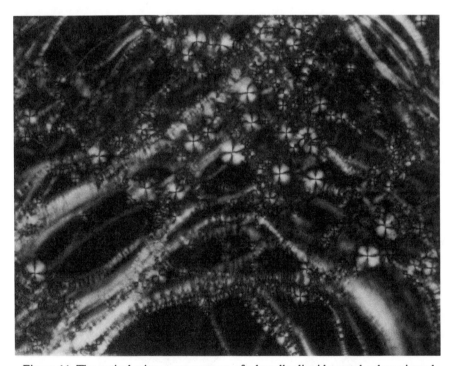

Figure 11. The optical microscopy pattern of a lamellar liquid crystal, when viewed between crossed polarizers.

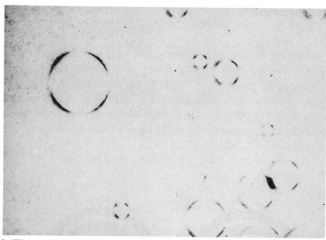

Figure 12. The radiant liquid crystal layers around the droplets is visible, when the emulsion is viewed between crossed polarizers.

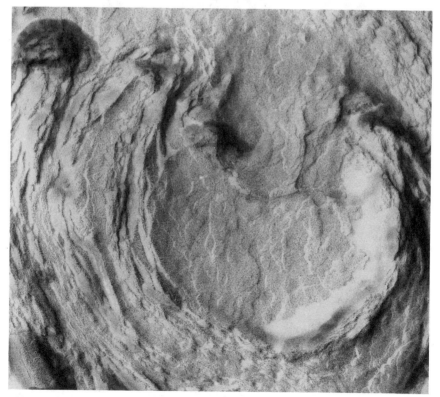

Figure 13. Electron microscopy of freeze fractured samples (15) show the individual layers of the adsorbed liquid crystal.

The reason the liquid crystal is adsorbed at the oil/water interface is because of its structure, Figure 3. At the interface the final layer towards the aqueous phase terminates with the polar group while the layer towards the oil finishes with the methyl layer. In this manner the interfacial free energy will be at a minimum.

At the interface the liquid crystal serves as a viscous barrier to accept and dissipate the energy of the flocculation, Figure 14 (24). This energy is the major part of the energy of the coalescence process leading to the final bilayer of the surfactant, Figure 15. The initiation of the flocculation process leads to very small energy changes, Figure 15 and good stability is assumed as long as the liquid crystal is adsorbed.

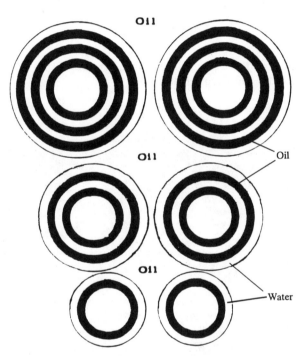

Figure 14. The coalescence process of a droplet covered with a lamellar liquid crystal consists of two stages. At the first the layers of the liquid crystal are removed two by two. The terminal step is the description of the final bilayer of the structure. (Reprinted with permission from Jansson, P. O.; Friberg, S. E. Mol. Cryst. Liq. Cryst. 1976, 34, 78. Copyright 1976 Gordon and Breach.)

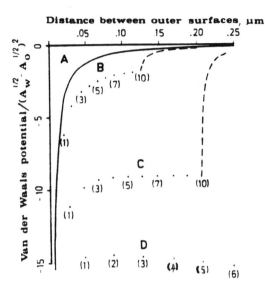

Figure 15. The driving energy for the initial
coalescence steps is extremely small. (Reprinted with
permission from Jansson, P. O.; Friberg, S. E. Mol.
Cryst. Liq. Cryst. 1976, 34, 75. Copyright 1976 Gordon
and Breach.)

Summary

The surfactant association structures are important both for microemulsions and emulsions, but different conformations are required to achieve efficient stabilization of the
two systems. The microemulsions require association with lowest energy at a very small
radius, while the emulsions demand emulsifier combinations, which favor a lamellar
packing.

Acknowledgment

This research was in part funded by an NSF grant CBT-8819140.

Literature Cited

1.   Food Emulsions; Friberg, S., Ed.; Marcel Dekker: New York, 1976.
2.   Dickenson, E.; Stanisby, G. Colloids in Foods; Applied Science: New York,
      1982.
3.   Lynch, M.N.; Griffin, W.C. In Emulsions and Emulsion Technology, Lissant,
      K.J., Ed.; Marcel Dekker: New York, 1974; Vol. 1, p 250.
4.   Darling, D.F.; Bickett, R.I. In Food Emulsions and Foams; Dickinson, E., Ed.;
      Royal Society of Chemistry: London, 1987; p 7.

5.   Grat, E.; Bauer, H. In Food Emulsions; Friberg, S.E., Ed.; Marcel Dekker: New York, 1976; p 295.
6.   Bracco, U.; Hidalgo, J; Bohren, H. J. Dairy Sci. 1972, 55, 165.
7.   King, N. The Milk Fat Globule Membrane and Some Associated Phenomena; Commonwealth Agricultural Bureaux: Formham Royal Bucks, 1955.
8.   Walstra, P; Oortwijas, H. Neth. Milk Dairy J. 1982, 36, 103.
9.   Tornberg, E. J. Sci. Food Agric. 1978, 29, 762.
10.   Izmailova, V.N. Progr. Surface Membrane Sci. 1979, 13, 141.
11.   Napper, D.N. Polymeric Stabilization of Colloidal Dispersions; Academic: New York, 1983.
12.   Pickering, S.U. J. Chem. Soc., 1934, 1112
13.   Friberg, S.E. In Food Emulsions; Larsson, K.; Friberg, S.E., Eds.; Marcel Dekker: New York, 1990; p 7.
14.   Pfüller, U. Mizellen-Vesikel-Mikroemulsioen; Springer Verlag: Berlin, 1986.
15.   Microemulsions: Structure and Dynamics, Friberg, S.E.; Bothorel, P., Eds.; CRC, 1987.
16.   Encyclopedia of Emulsion Technology, Becher, P., Ed.; Marcel Dekker: New York, 1987; Vol. 2.
17.   Larsson, K. J. Disp. Sci. Technol. 1980, 1, 267.
18.   Napper, D.N. Polymeric Stabilization of Colloidal Dispersions; Academic: New York, 1983.
19.   Friberg, S.E.; Rydhag, L. J. Am. Oil Chem. Soc. 1971, 48, 113.
20.   Lawrence, A.C.S.; Pearson, J.T. Proc. Int. Congr. Surf. Act. Subst. 1964, 2, 709.
21.   Ekwall, P. Advances in Liquid Crystals, Brown, G.H., Ed.; Academic: New York, 1975; Vol. 1, p 1.
22.   Krog, N. In Food Emulsions; Friberg, S.E., Ed.; Marcel Dekker: New York, 1976; p 67.
23.   Barry, B.W. Advanced Colloid Interface Sci. 1975, 5, 37.
24.   Jansson, P.O.; Friberg, S.E. Mol. Cryst. Liq. Cryst. 1976, 34, 75.
25.   Friberg, S.E.; Jansson, P.O.; Cederberg, E. J. Colloid Interface Sci. 1976, 55, 614.

RECEIVED August 16, 1990

MICROEMULSIONS IN FOODS

Chapter 3

Solubilization of Water and Water-Soluble Compounds in Triglycerides

Magda El-Nokaly, George Hiler, Sr., and Joseph McGrady

Food and Beverage Technology Division, Miami Valley Laboratories, The Procter and Gamble Company, P.O. Box 398707, Cincinnati, OH 45239–8707

A water-in-oil (W/O) microemulsion was chosen as the delivery system for water soluble nutrients, and flavors in foods. An ingestible, co-surfactant free system, with no off-taste or change in performance is described. Commercially available food surfactants are evaluated based on their structures and performance in solubilizing water in the high triglyceride concentration range (above 90%). The conditions needed to form such a microemulsion with minimum surfactant amounts are discussed.

Microemulsions were considered in an effort to develop the capability to deliver water-soluble nutrients, flavors, and flavor enhancers to foods through a lipid based system. This paper describes our successful attempt to formulate a good tasting and performing water-in-oil microemulsion, containing a minimum amount of surfactant using commercially available surfactants.

A generally accepted definition of a microemulsion is: "A clear, thermodynamically stable homogeneous dispersion of two immiscible liquids containing appropriate amounts of surfactants and co-surfactant". Friberg points out a great number of systems are not thermodynamically stable and a change of the definition requiring spontaneous formation would be more suitable (1).

The term "microemulsion" has been used to describe different types of solutions (2):

A substance which is otherwise insoluble in the bulk phase, may form a molecular solution upon addition of a third solvent component. Such systems of enhanced solubility by mixing solvents are often termed "detergentless microemulsions" (Figure 1a).

Addition of a proper surfactant (SAA) with or without a co-surfactant (COS), will form a clear solution of the insoluble substance. It can be a co-solubilization in which the surfactant and co-surfactant form a liquid which can dissolve both oil or water as a molecular solution (Figure 1b).

Another possibility is the formation of large and well defined surfactant aggregates, with or without co-surfactants.

0097–6156/91/0448–0026$06.00/0

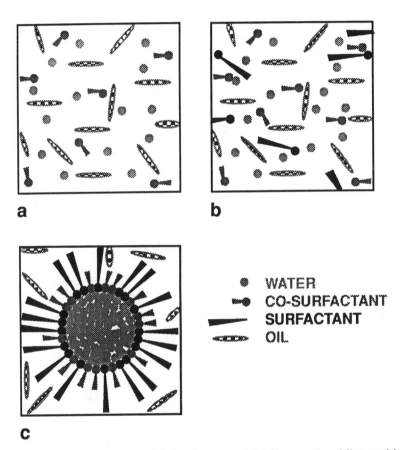

Figure 1: Schematic Illustrations of the Structure of (a) Detergentless Microemulsion (Molecular Solution); (b) Co-Solubilization (Molecular Solution); and (c) Microemulsion (Water-in-Oil).

The core of such aggregates dissolves the insoluble in the bulk substances (Figure 1c). Such systems have been discussed in other Chapters of this book and in the literature (3-5).

There are very few available examples of ingestible water-in-oil or oil-in-water microemulsion systems for food applications, even though much has been accomplished in recent years in the general field of microemulsions.

Earlier Work on Food Microemulsions Applications

Examples of Oil-in-Water Applications. Reports on milk fortified with vitamin A solubilized in a microemulsion have been recently published (6).

Dissolving essential oils in water with and without an alcohol co-surfactant for aromatization of beverage or pharmaceutical formulations, have been reported (7-9).

Figure 2 shows the microemulsion single phase region of a peppermint oil/Tween 20/water system studied by Treptow (9).

Friberg solubilized up to 15 wt% tricaprylin in an isotropic aqueous solution of monocaprylin and a hydrotrope such as sodium xylene sulfonate (10)

Treptow dissolved less than 10% soybean oil in water with a 30/70 surfactant mixture of Tween 20 and G1045 respectively (ICI, Inc.). This mixture had an HLB of 13.1 (9).

Examples of Water-in-Oil Applications. The literature available on water-in-oil, the oil being triglycerides, microemulsion is reviewed in Table I. With the exception of the work done by Larsson (12) and Treptow (9), an alcohol or acid co-surfactant was needed to form the microemulsion. Such co-surfactants are usually not acceptable for taste, safety, or performance reasons in ingestible oil formulations.

Examples of non-ingestible water-in-triglycerides are more numerous. They have gained considerable attention recently due to the applicability of these oils in a number of new ways, such as engine fuels, lubricating oils, and carrier liquids (14,15).

The sparcity of work available on water solubilization in triglyceride for food applications is obviously due to difficulties inherent in the structure of oil, in finding appropriate co-surfactants, and the need to use minimum amounts of food approved surfactants (emulsifiers) not to adversely affect the oil properties.

The paragraphs below describe our efforts to better understand what is needed to be taken into consideration to form edible water-in-triglyceride microemulsions.

Characteristics of Microemulsions

Microemulsions of the aggregate type (Figure 1c) possess special characteristics of relatively large interfacial area, ultra low interfacial tension, and large solubility capacity as compared to many other colloidal systems.

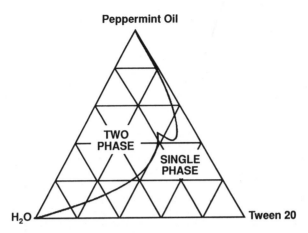

Figure 2: Phase Diagram: Peppermint Oil/Water/Tween 20 at 25 °C Showing the Single Microemulsion Phase Region. (Reprinted from ref. 9.)

TABLE I. EARLIER WORK
EXAMPLES OF WATER-IN-TRIGLYCERIDES

Oil	Surfactant	Co-Surfactant	References
Soybean oil	POE Sorbitol oleate	---	9
Tricaprylin	Alkyl aryl polyglycol ether	Pentanol	10
Trioctanoin	Sodium octanoate (soap)	Octanoic acid	11
Soybean oil	Monoglyceride	---	12
1,2,3-[Tris (2 ethylhexa- noyloxy)] propane	Tetraethylene glycol dodecyl ether $(R_{12}EO_4)$	Hexadecane	13
Canola oil	Monoglyceride Acetic acid ester of monoglyceride	Tert-butanol Isopropanol 1-Hexanol	14
Triglyceride	None*	Ethanol 1-Butanol	15
Soybean oil	0-Alkyl-3 D-glucose	Ethanol	16

*Detergentless microemulsion

The conditions necessary for microemulsion formation are:
- Large adsorption of SAA or SAA/COS mixture at the W/O interface achieved by choosing a SAA mixture with proper HLB.
- High fluidity of the interface. The interfacial fluidity can be enhanced by using a proper co-surfactant or an optimum temperature.
- Optimum curvature. The importance of oil penetration in the SAA/COS interface and the appropriate SAA/COS structures (17).

Importance of Bulk Oil Properties. Why is it difficult to form a triglyceride microemulsion?

Triglycerides are semi-polar compared to hydrocarbons. A surfactant of higher HLB is thus needed to favor the water-in-oil system, with lower solubility in bulk and increased adsorption at the interface. In the case of triglyceride, the ratio of SAA/W is large. The emulsifier efficiency is decreased if it is lost to the bulk and is unavailable to the interface. Treptow found a maximum normal (L_1) and reverse (L_2) micellar regions (3), were formed with a surfactant HLB of 13.1 and 9.2 respectively (9).

Edible triglycerides such as soybean, rapeseed, or sunflower oils contain long alkyl chains mainly C_{16}, C_{18}, C_{20}, and C_{22}. The oil may be too bulky to penetrate the interfacial film to assist the formation of the optimum curvature (Figure 3). Reports of oil being solubilized in the aggregates palisade layer may be due to the shortness of the alkyl chains in the triglycerides used (13).

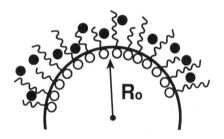

〜〜O **Surfactant**
〜O **Co-Surfactant**
● **Oil**

Figure 3: Optimum Curvature, Ro=Radius of Spontaneous Curvatures.

Role of a Co-Surfactant. A co-surfactant is usually a medium
chain fatty alcohol, acid, or amines (18,19). It is usually
chosen to be widely different in hydrocarbon moiety size compared
to the surfactant. The role of the co-surfactant together with
the surfactant is to lower the interfacial tension down to a very
small even transient negative value at which the interface would
expand to form fine dispersed droplets, and subsequently adsorb
more SAA and SAA/COS until their bulk condition is depleted
enough to make interfacial tension positive again. This process
known as "spontaneous emulsification", forms the microemulsion.
Thus, based on ability of the co-surfactant to affect the solvent
properties of oil and/or water and to penetrate the surfactant
interfacial monolayer, it can:
· Decrease further the interfacial tension. Increase the
  fluidity of interfaces
· Destroy liquid crystalline and/or gel structures which prevent
  the formation of microemulsions
· Adjust HLB value and spontaneous curvature of the interface by
  changing surfactant partitioning characteristics
· Decrease the sensitivity to composition fluctuations and
  brings formulation to its optimum state
     The presence of a co-surfactant in triglyceride increases
the microemulsions area (10, 11, 13-16).

Nature and Concentration of Surfactant. The conditions
surrounding our application are: Co-surfactants are not easy to
find in foods and their addition is not a theoretical
requirement. The oil structure and properties are constant.
The formulated water-in-triglyceride microemulsion are to be
shelf stable at room temperature. The temperature could not be
varied to favor microemulsion formation. Thus, the nature and
concentration of the surfactant become of utmost importance to
obtain a maximum solubilization in a given W/O microemulsion.
The natural radius and fluidity of the interface should be
adjusted to optimal values at which the bending stress and the
attractive force of the interfaces are both minimized.
     The concepts to be considered when designing COS-free W/O
microemulsions were reported to be (20):
· Near equal partitioning of surfactant between the liquid
  phases.
· Maximal linear extension of hydrophobic or hydrophilic end of
  the molecule or both.
· Forming a fluid interfacial film using double or branched
  chain surfactant at a temperature above the thermotropic phase
  transition temperature by weakening surfactant lateral
  interaction.
· It is necessary but not a sufficient condition that the
  surfactant hydrocarbon volume (v), effective chain length
  (L_c), and head group (a_o) should satisfy the relation $v/a_o L_c$
  > 1 (W/O) (21).
     Thus, a surfactant with a laterally bulky hydrocarbon part
on a relatively small head group, such as some of the double
chain surfactant, favors the formation of W/O aggregates.

32 MICROEMULSIONS AND EMULSIONS IN FOODS

A number of double-tailed ionic surfactants such as Aerosol OT are known to form co-surfactant-free microemulsions. Abe et. al. (20), have shown that a spectrum of ethoxylated double branched tail alkane sulfonates have the ability to microemulsify hydrocarbon and electrolyte without added COS. The optimal structures were obtained through parallel optimization of:
· Hydrophobe branching
· Surfactant molecular weight including both hydrophobe and hydrophile
· Surfactant head group

Food Emulsifiers or Surfactants

Surfactants in food are usually called emulsifiers whether their intended use is emulsification or not. An Acceptable Daily Intake (ADI) value has been allocated to most food emulsifiers by health authorities in many countries (FDA, FAO, EEC) (Table II).
Chemically, most food surfactants are esters of fatty acids with naturally occurring alcohols and acids. The primary raw materials for food surfactants production are fats and oils which can be utilized directly or after having been hydrogenated, fractionated, or split to fatty acids and glycerol (22). They have to impart no taste or smell on foods.
They are mostly nonionic surfactants with few exceptions such as succinic, citric, and diacetyl tartaric acids esters of monoglycerides and soaps. Amphoteric lecithin is the only food approved surfactant containing a positive charge.

MATERIALS AND METHODS

Crisco Oil: Soybean oil

Surfactants: Many food surfactants were tested some are reported in Table III a and b. The following are the ones found most efficient:
· Polyglycerol oleate (AM #506, Grindsted) based on esterification with oleic acid (min. 92%).
· Polyglycerol linoleate (AM #507, Grindsted) based on esterification with sunflower oil fatty acids.
The average composition, of the polyglycerol moiety in these esters are 18 di-, 18 tri-, 31 tetra-, 13 penta-, 10 hexa-, 2% hepta and the remaining 8% is glycerol and higher glycerol. The FA composition is 36 mono-, and 63 di-.
· Monoglyceride (AM #505, Grindsted) is Dimodan LS with no saturated fatty acid chains.
· Atlas G1186, Polyoxyethylene sorbitol oleate, (ICI).

Preparation of Surfactants. Polyglycerol linoleate (AM #507), was not completely soluble in Crisco oil as such, unless water was added to form the microemulsion.

The Crisco oil/water/AM #507 phase diagram in Figure 4, was constructed using the surfactant samples:

· Sample - 1, AM #507 as provided by Grindsted.
· Sample - 2, AM #507 purified as follows to remove polyglycerol.

70 grams of polyglycerollinoleate and 350 grams of ethyl acetate were placed in a large separatory funnel. 1500 milliliters of hot (75°C) water was then added and the mixture shaken well. It was left standing overnight. The ethyl acetate layer was separated and dried over magnesium sulfate. The ethyl acetate was evaporated in a rotary evaporator followed by eight hours on a vacuum pump, 65.427 grams were recovered.

Preparation of Phase Diagram. A Zymark Pye Robotic system, Figure 5, described elsewhere (23) was used to prepare samples to construct phase diagrams of the various surfactant/Crisco oil/water systems tested. In some cases, the whole phase diagram was constructed, but generally only the region below 12 wt% surfactant was screened. The samples were left to equilibrate in 75 and 100°F storage rooms.

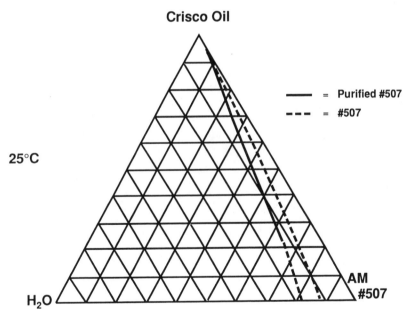

Figure 4: Phase Diagram: Crisco Oil/Water/Polyglycerol
Linoleate
- - - - AM#507 (Sample 1)
———— AM#507 Extracted with Ethyl Acetate (Sample 2)
(The lower part of the curve (---) was not done)

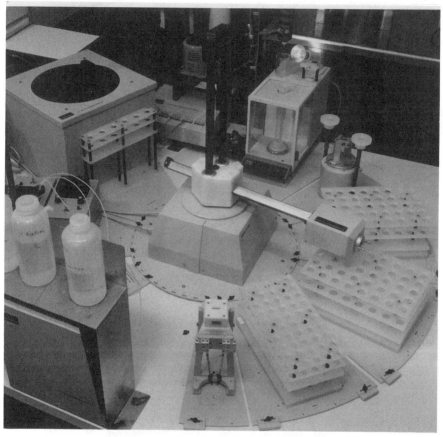

Figure 5:   Zymark Pye Robotic System

RESULTS AND DISCUSSION

Effect of Surfactant Structures

Surfactants chosen from the different classes previously listed in Table II, were tested for water-in-oil microemulsion formation at 12:1, SAA/W by weight.

Partial Glycerides. (Table III A) Varying the glycerides fatty acid composition, as seen in Table III A, showed Dimodan LS (12) and AM #505 (Dimodan LS with no saturated fatty acid chains), to give the lowest surfactant to water ratio 10:1 wt/wt. Dimodan LS is also the highest in unsaturation compared to other glycerides tested, ~ 80% in C18:2. Nevertheless, to dissolve 0.1% water, a minimum amount of 4% surfactant was needed.
    Dimodan LS could not be chosen because of the relative high SAA/W ratio needed to form a microemulsion and an objectionable coating of the mouth and after taste. Removing its saturated fatty acid components as in AM#505 improved the mouthfeel, but not the SAA/W ratio.

Fatty Acid Esters of Polyols. Surfactants containing varied hydrophilic groups such as sorbitol, sorbitan, or polyglycerol esters with different glycerol units were tested to study their effect on SAA/W ratio.
    Table III B shows the effect of varying the number of glycerol units in commercial polyglycerol esters. The oleate and linoleate groups were preferred for testing because of their unsaturation and its known effect in increasing fluidity of interfacial film. Some laurates were tested unsuccessfully.
    The optimum number of glycerol units to have in the molecule for a microemulsion at SAA/W below 12:1, was four as seen in Table III B for Polyaldo 4-2-0 and AM#507.

Another requirement seems to be the presence of large amounts of di-fatty acid chains mixed with mono. AM#507 contains 63% dilinoleate and 36% mono.
    The sorbitan and sorbitol esters of oleate and linoleate did not give microemulsions at SAA/W ratio below 12:1.

Crisco Oil/Water/AM#507 Phase Diagrams

Extraction of the polyglycerol linoleate (AM#507) with ethyl acetate removed the free unreacted polyglycerol. The microemulsion region formed by the sample-1, as provided by Grindsted, was smaller than that of the purified sample-2, as seen in Figure 4. Except for the upper oil corner, above 90%, where they behaved similarly. This being the area of interest for our study, we did not find it necessary to work further with purified samples.

TABLE II A CHEMICAL CLASSFICATION OF
FOOD EMULSIFIERS AND LEGAL STATUS (US FDA 21 CFR*)

General Class	Example	ADI Values** mg/kg body wt/day
Partial Glycerides	Mono- and diglycerides	Not limited
	Lactic acid esters of monoglycerides	Not limited
	Acetic acid esters of monoglycerides	Not limited
	Citric acid esters of monoglycerides	Not limited
	Diacetyl tartaric acid esters of monoglycerides	50
	Succinic acid esters of monoglycerides	3% by wt. of shortening; 5% by wt. of flour
Fatty Acid Esters of Polyols	Propylene glycol esters of fatty acids	25
	Polyglycerol esters of fatty acids	25
	Sucrose esters of fatty acids	10
	Sorbitan esters	25
Ethoxylated Emulsifiers	Ethoxylated partial glycerides	--
	Polysorbates	25
Phosphatides, Phosphorylated Partial Glycerides	Lecithin	Not limited
	Fractioned phosphatides	--
	Phosphorylated monoglyceride	--
Miscellaneous	Sodium stearoyl-lactylate	20
	Caldium stearoyl-lactylate	20
	Salts of fatty acids (Na,K)	Not limited
	Sodium dodecyl sulfate	

*CFR Code of Federal Regulations
**Acceptable Daily Intake (ADI) for man

TABLE III SURFACTANTS FORMING W/O MICROEMULSION
IN CRISCO OIL AT S/W = 12 OR LESS

A. GLYCERIDES

Surfactants Trade Name	Composition	S/W wt/wt
Dimodan LS (G)	Sunflower (80% C18:2; 20% C18:1)	10:1*
AM #505 (G)	No saturated FA	10:1*
Dimodan O (G)	Soybean oil (51% C18:2; 23% C18:1)	None
AM #489 (G)	Rapeseed oil (40% di-) (21% C18:2; 62% C18:1)	None
AM #490 (G)	Sunflower Oil (39% di-)	None
Monooleate (P)	Monooleate	None
Aldo MD (L)	Mono- and dioleate	None
Aldo MLD (L)	Monolaurate	None

B. POLYGLYCEROL ESTERS OF FATTY ACIDS

Surfactant Trade Name	Composition	S/W wt/wt
Caprol 3GO (C)	Triglycerol monoleate .	None
Polyaldo 4-2-0 (L)	Tetraglycerol dioleate	10:0.71
Caprol 6G20 (C)	Hexaglycerol dioleate	None
Polyaldo 2010 (L)	Decaglycerol dioleate	None
AM #506 (G)	Polyglycerol oleate**	None
Experimental (L)	Tetraglyceryl laurate	None
Polyaldo 4-2-L (L)	Tetraglyceryl dilaurate	None
93-919 (L)	Polyglycerol monolaurate	None
AM #507 (G)	Polyglycerol mono-dilinoleate**	9:1
Triodan 20 (G)	Polyglycerol esters	14:1
Homodan PT (G)	Polyglycerol ester of dimerized soybean oil	None

*Microemulsion formed
**Polyglycerol: mainly tetra
  G = Grindsted
  L = Lonza/Glyco
  C = Capitol City
  P = Pflatz & Baur

Comparison of Microemulsion Region in Phase Diagram: Crisco
Oil/Water/Emulsifier for Different Emulsifiers. The surfactants
(emulsifiers) used to form the phase diagram in Figure 6 are the
polyglycerol linoleate (AM#507), oleate (AM#506), monoglyceride
(AM#505), and polyoxyethylene sorbitol oleate (Atlas G1186). The
surfactants behave differently in the upper oil and lower
surfactant corners. Polyglycerol oleate forms the smaller
microemulsion region throughout the whole range. In the lower
50% towards the surfactant corner, the POE sorbitol oleate and
monoglyceride form the larger microemulsion area. The
monoglyceride shows the typical knee, indicating the presence of
liquid crystals. In the upper oil corner, above 90%, the
polyglycerol linoleate solubilizes more water at lower surfactant
concentration as seen in Figure 7.
    Table IV shows the ratio of surfactant to water for the
different food emulsifiers used. For every molecule of
polyglycerol linoleate (AM#507) four molecules of water are
solubilized.
    The rest of the phase diagram Crisco oil/water/AM#507, was
scanned for a one phase oil-in-water microemulsion or liquid
crystalline regions. None were found instead mixture of two and
three phases of oil, water, and liquid crystals.

Structure of Crisco Oil/Water/AM#507 Microemulsion. The
surfactant polyglycerol linoleate (AM# 507) structure and
composition were designed to give optimum conditions for a
COS-free W/O microemulsion and they did.
    The concepts earlier described (20), which should give
spontaneous curvature, increase the interface fluidity, and
decrease further the interfacial tension were followed as
practically possible. The polyglycerol linoleate is a long chain
surfactant affording a maximal linear extension of the
hydrophobic and hydrophilic end of the molecule. It may even be
worth mentioning that the length of the extended hydrophobic and
hydrophilic groups, as measured from the carboxy group, is almost
equal, 19.148A and 19.75A, for the monolinoleate and the
tetraglycerol respectively (Figure 8). The surfactant is a
mixture of 63% branched dilinoleate, the rest straight
monolinoleate with some oleate. Branching should be another
factor in increasing the interfacial fluidity. Figure 8 shows
the special kink in the linoleate molecule which makes its
alignment difficult, thus leading to an increase in the
interfacial fluidity. The presence of unsaturation was found to
be important.

The tetraglycerol oleate did no give as good a SAA/W ratio as the
linoleate. Those attributes were chosen to lower the
thermotropic phase transition temperature by weakening surfactant
lateral interaction. The monooleate and linoleate could act as
co-surfactant since they are different in hydrocarbon moiety size
compared to the dilinoleate. The hydrophile group size giving
the lowest SAA/W ratio was the tetraglycerol (Table III B).
    Attempts were made at studying the structure of the
microemulsion formed by the polyglycerol linoleate (AM #507),
water and Crisco oil. The results were inconclusive as to the

Soybean Oil

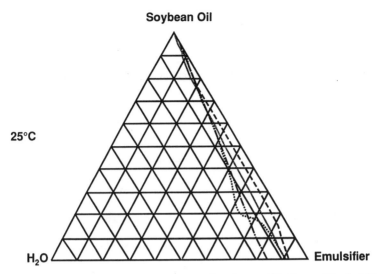

25°C

H₂O — Emulsifier

Figure 6: Complete Phase Diagram of Crisco Oil/Water/Emulsifier
Emulsifiers   AM#507 Polyglycerol Linoleate (sample 1)
                AM#506 Polyglycerol Oleate
                AM#505 Monoglyceride
                Atlas G1186 Polyoxyethylene Sobitol
                Oleate

Crisco Oil

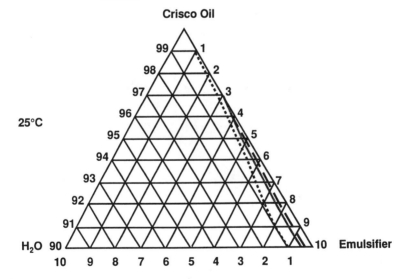

25°C

H₂O 90 — 10 Emulsifier

Figure 7: The Upper Oil Corner (Above 90%) of the Phase
Diagram Crisco Oil/Water/Emulsifier
Emulsifiers ■■■■ AM#507 Polyglycerol Linoleate (sample 1)
            ■ ■ ■ AM#506 Polyglycerol Oleate
            ⅰⅰⅰⅰⅰⅰⅰⅰ AM#505 Monoglyceride

TABLE IV RATIO OF SURFACTANT TO WATER FOR DIFFERENT FOOD EMULSIFERS

Emulsifier	SAA/W moles/moles
AM #505	~1.83
AM #506	<0.83
AM #507	<0.23

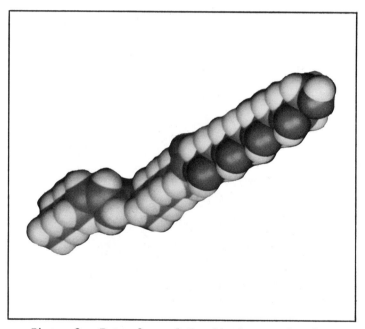

Figure 8:   Tetraglycerol Monolinoleate Molecule

presence of well defined aggregates. Thus, the possibility exist of a surfactant-water co-solubilization system of the type earlier discussed and shown in Figure 1b.

The system may be going from a co-solubilized directly to liquid crystalline because of the surfactant strong tendency to form liquid crystals (Lindman, B., University of Lund, Sweden, personal communication, 1990).

Solubilization of Water Soluble Flavors in Crisco Oil/Water/ Polyglycerol/Linoleate

The water soluble flavors and flavor enhancers we solubilized in the water-in-Crisco oil microemulsion are shown in Table V. The reported concentration in ppm is that needed to deliver the

TABLE V WATER SOLUBLE FLAVORS DELIVERED BY CRISCO OIL/WATER/POLYGLYCEROL LINOLEATE MICROEMULSION

Water Soluble Flavor	Amount in Oil (ppm)
Ribotide	1000
Furaneol	10
Cyclotene	100
Methionine sulfoxide	100
Caramel furanone	100
S-methyl methionine	100
3-hydroxy-4,5-dimethyl-2(5H)-furanone	100
Natural butter flavor	1000
Honey flavor	1000
Fructose	5000

Others

Enzymes, sugars, salts, citric, or ascorbic acids

flavor to oil and french fries during frying with the
microemulsion, and not the amount of compound necessary to
saturate the water in the water-in-oil microemulsions. This
study will not be reported in more detail at this
time.

SUMMARY

Various commercially available food emulsifiers were tested to
form an alcohol or acid-free water-in-triglyceride microemulsion.
The surfactant-to-water ratio was smallest for a polyglycerol
linoleate at low water content. Its hydrophobe was unsaturated,
mainly linoleate (C18:2) with some oleate (C18:1), and branched
(63% dilinoleate) mixed with straight chains. The hydrophile, to
give optimum solubilization, was the tetraglycerol.
    The microemulsion was good tasting, low foaming, and
non-spattering during frying. It solubilized various water
soluble flavors and nutrients which could delivered to a variety
of food products.

ACKNOWLEDGMENTS

The Grindsted Company, Braband, Denmark synthesized the
experimental surfactants and provided for their analysis. Mrs.
Carla Cobb typed the manuscript.

LITERATURE CITED

1. Friberg, S. E. Colloids and Surfaces 1982, 4, 201
2. Shah, D. O.; Walker, R. D.; Hsieh, W. C.; Shah, N. J.;
   Dwivedi, S.; Nelander, J.; Pepinsky, R.; Deamer, D. W. "Some
   Structural Aspects of Microemulsions and Co-Solubilized
   Systems"; SPE 5815 paper presented in Tulsa, OK, March 22-24,
   1976.
3. Friberg, S. E. J. Amer. Oil Chem. Soc. 1971, 48, 578-81.
4. Shinoda, K.; Lindman, B. Langmuir 1987, 3, 135-49.
5. Lindman, B.; Shinoda, K.; Olsson, U.; Anderson, D.; Karlström,
   G.; Wennerström, H. Colloids and Surfaces 1989, 38, 205-24.
6. Duxbury, D.D. Food Processing May 1988, 62-4
7. Wolf, P.A.; Havekotte, M. J. US Patent 4 835 002, 1989.
8. Thoma, K.; Pfaff, G. Perfumer and Flavorist 1978, 2, 27.
9. Treptow, R. S. Research and Development Report, June 1, 1971;
   The Procter and Gamble Company, Cincinnati, OH.
10. Friberg, S. E.; Rydhag, L. J. Amer. Oil Chem. Soc. 1971, 48
    (3), 113-15.
11. Friberg, S. E.; Gezelius, L. H.; Wilton, I. Chem. Phys. Lipids
    1971, 6, 364-72.
12. Gulik-Krzywicki, T.; Larsson, K. Chem. Phys. Lipids 1984, 35,
    127.
13. Kunieda, H.; Asaoka, H.; Shinoda, K. J. Phys. Chem. 1988, 92,
    185.
14. Vesala, A. M.; Rosenholm, J. B.; Laiho, S. J. Amer. Oil Chem.
    Soc. 1985, 62 (9), 1379.

15. Schwab, A. W.; Nielsen, H. C.; Brooks, D. D.; Pryde, E. H. J. Dispersion Sci. Technology. 1983, 4 (1), 1-17.
16. Chelle, F.; Ronco, G. L.; Villa, P. J. Patent W088/08000, 1988.
17. Bansal, V. K.; Shah, D. O.; O'Connell, J. P. J. Colloid Interface Sci. 1980, 75 (2), 462-75.
18. Stilbs, P.; Rapacki, K.; Lindman, B. J. Colloid Interface Sci. 1983, 95, 583-85.
19. Lang, J.; Rueff, R.; Dinh-Cao, M.; Zana, R. J. Colloid Interface Sci. 1984, 101 (1), 184-200.
20. Abe, M.; Schechter, D.; Schechter, R. S.; Wade, W. H.; Weerasooriya, U.; Yiv, S. J. Colloid and Interface Sci. 1986, 114 (2), 342-56.
21. Mitchell, D. J.; Ninham, B. W. J. Chem. Soc., Faraday Trans. 2 1981, 77, 601-29.
22. Lauridsen, J.B. Food Surfactants, Their Structure and Polymophism; Technical Paper No TP 908-1e, 1986; Grindsted Products A/S, Braband, Denmark.
23. Osborne, D. W.; Pesheck, C. V.; O'Neill, K. J.; Ward, A. J. I.; El-Nokaly, M. Cosmetics and Toiletries 1988, 103, 53-62

RECEIVED September 9, 1990

Chapter 4

Emulsions of Reversed Micellar Phases and Aqueous Dispersions of Cubic Phases of Lipids

Some Food Aspects

K. Larsson

Department of Food Technology, University of Lund, P.O. Box 124, S–221 00 Lund, Sweden

The L2- and cubic phases formed by food lipids can be emulsified or dispersed in water. The L2-phase will then give a water-in-oil-in-water structure and the cubic phase represents a three-dimensionally packed membrane with water on each side. The L2-phase is very efficient in order to provide lipid oxidation protection, whereas the cubic phase is able to incorporate and stabilize enzymes, and such application possibilities are demonstrated.

Among aqueous dispersions of lipids, emulsions of triglycerides (oils/fats) and liposomes of polar lipids are best known. The technology of emulsification of oils has reached a high level of sophistication with regard to processing technique as well as use of various emulsifiers. Less is known concerning dispersion of the lamellar liquid-crystalline phase (L_α) in excess of water into a liposomal dispersion. One obvious requirement is that the lamellar phase is not allowed to exhibit "infinite" swelling, and therefore anionic or cationic lipids cannot be used. It should be pointed out that liposomes in this context are defined as multilayer particles contrary to vesicles, which are closed unilayers.

There are three other types of lipid phases (beside the crystalline state), which can coexist in equilibrium with excess of water and thus can form aqueous dispersions. Two are liquid-crystalline; the reversed hexagonal (H_{II}) and the cubic types. The third is the reversed micellar phase, sometimes microemulsion type, which is termed L2-phase. Two of these phases are particularly interesting from an application point of view, the dispersion of cubic phases and emulsions of L2-phases. This

0097–6156/91/0448–0044$06.00/0

paper is mainly focused on the preparation of such phases from food-grade lipids. Some applications possibilities are finally discussed.

Phase Diagrams Used for L2- and Cubic Phase Dispersions

Monoglycerides are a dominating group of functional additives used in the food industry, and usually the effect is due to the actual phases formed with water whereas the separate monomers in the aqueous medium are not involved (one important exception is starch interaction). We will consider the sunflower oil monoglycerides-water system in detail, see Fig. 1, as it exhibits the phases used in our dispersion studies. At room temperature the addition of water to monoolein results in swelling of the L2-phase, and then two cubic phases are formed. At excess of water the very viscous cubic phase coexists with a solution with very low monomeric concentration (about 10^{-6} M). When such a phase is heated, it is transformed *via* the reversed hexagonal phase into the L2-phase. This liquid, containing water aggregates as discussed below, can also coexist with excess of water.

If we want to obtain an L2-phase of edible lipids at room temperature, one way is to add a triglyceride oil to the monoglyceride-water mixtures discussed above. Such a system, formed by soybean oil, the same sunflower oil monoglycerides as discussed above (Dimodan LS from Grindsted Products) and water is shown in Fig, 2 (1,3). There is a limit of association of the water molecules at about 10% (w/w) and above this concentration there is a successive increase in water swelling towards a maximum of about 15% (w/w), with increasing amount of monoglyceride.

Structure of the L2-Phase

The L2-phase as shown in Fig. 2 could be classified as a microemulsion as it contains oil/water/surfactant in a thermodynamically stable form. A remarkable structural feature on such lipid L2-phases is that it exhibits X-ray diffraction characteristics closely related to the corresponding liquid-crystalline phase which can be obtained on cooling (4). In the particular case of monoglyceride-water systems a lamellar type of structure lacking long-range order was therefore proposed (4). The structure as evident from a freeze-fractured sample is shown in Fig. 3 together with the lamellar liquid-crystalline phase. The proposed structure based on X-ray data is clearly confirmed, bundles of lipid bilayer units in a disordered array can thus be seen.

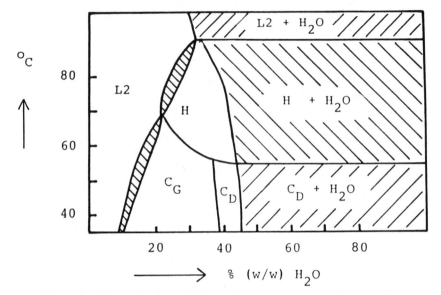

Figure 1.   Phase diagram of the aqueous system sunflower oil monoglycerides. The hexagonal region is denoted H̲, and the two cubic phases are called C̲D and C̲G (corresponding to the diamond and gyroid structure types).

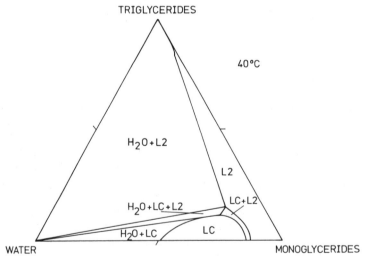

Figure 2.   Phase diagram of a ternary system monoglycerides-triglycerides-water at 40°C. The monoglyceride is the same as that used in Fig. 1, and the triglyceride component is soybean oil (1).

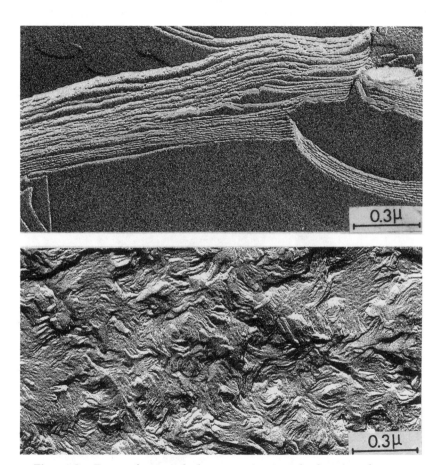

Figure 3.   Freeze-fractured electron micrograph showing the structure of an L2-phase (below) compared to that of an $\underline{L}_\alpha$-phase (above).

Structure of the Cubic Phase

Cubic phases are found in aqueous systems of most types of polar lipids, such as monoglycerides, phospholipids and galactolipids. They have a remarkable structure which best can be described as infinite periodic minimal surfaces (2). Such a surface has zero average curvature everywhere, *i.e.* it is as convex as it is concave in all positions on the surface. Such periodic surfaces without intersections were described in mathematics already during the last century. From X-ray space-group symmetry, electron microscopy data, NMR diffusion data etc. it is now clear that the cubic phases of lipids are ordered in just this way. A plastic model of such a surface is shown in Fig. 4. The lipid bilayer is centred on this surface and on both sides there are continuous water channel systems.

Figure 4. Structure of the cubic structure of the diamond type occurring in aqueous monoglyceride-water systems. The plastic model shows the periodic minimal surface structure located in the middle of the lipid bilayer.

Emulsification of L2-Phases

We have in our laboratory studied the interfacial tension between water and L2-phases of the type shown in Fig. 2. The value is quite low; about 1.5 mN/m (5). In order to produce stable emulsions, various proteins were tried. The best result was obtained by sodium caseinate. Even if the additional reduction in surface tension was not very high, the final value was about 1 mN/m, the emulsions obtained were quite stable. Obviously this protein is irreversibly adsorbed at the interface with many segments. With a concentration of 1% protein in the outside water medium quite stable emulsions of about 10% (w/w) L2-phase in water was obtained. Such an emulsion can also be described as a w/o/w-emulsion.

Dispersion of the Cubic Phase

If the cubic phase is stirred in excess of water, a kinetically stable dispersion is obtained under conditions when there is an equilibrium between the cubic phase, the lamellar phase and water. Such as equilibrium can be obtained in the ternary system of bile salts/monoolein/water or in phosphatidyl choline/monoolein/water (2). The stability of these dispersions is comparable to liposomal dispersions, and they can also be obtained in a similar size distribution, related to the mechanical energy input.

From X-ray and optical studies of such dispersions it is clear that the core has a cubic structure with its outer bilayer fused with the spherical inner bilayer of a liposome (2). These particles were therefore termed "cubosomes" (2). In the extreme case there might be only one such outer layer. Processing aspects on the dispersion technology is further discussed below.

Application of the L2-Phase

There are obvious advantages of having one stable phase which can solve hydrophilic as well as lipophilic substances, for example in connection with flavouring. We have been particularly interested in the use of L2-phases in order to protect lipids against oxidation. Our results from these studies will be summarized below, and details of the work is given in Ref. (6).

In order to simulate the lipid oxidation protection in a living system such as the cell membrane, it is necessary to keep tocopherol and ascorbic acid in molecular contact. The L2-phase provides such a possibility. In our studies we used an L2-phase able to solubilize 1% (w/w) of water (see Fig. 2), and in that water 5% (w/w) of ascorbic acid was solved. In the oil a concentration of 200 ppm tocopherol was used. The antioxidant effect obtained in such a system is drastic compared to earlier reported effects. The oil was stored in open glass vessels at 40°C in order to accelerate oxidation. Without antioxidants rancidity flavour was detected after a few days, and after a couple of weeks the oil started to polymerize. With this molecularly dispersed ascorbic acid-tocopherol system, however, the oil could be kept up to 100 days with only minor oxidation changes (the oxidation was followed by COP-analysis).

Application Possibilities of the Cubic Phase

Dispersions of cubic lipid-water phases using a valve homogenizer have been prepared by Tomas Landh in our department. He has been able to obtain dispersions with as high kinetic stability as liposomal dispersions in a size range around 10 μ in diameter by pluronics as stabilizers. Such dispersions can be introduced in water-continuous foods, with enzymes localized in and stabilized by the cubic phase. Examples of interesting enzymes are superoxide dismuthase as antioxidant and lysozyme as antimicrobial protection agents. The lysozyme is protected within the phase and slowly released into the water environment.

Another application possibility is to introduce glycose oxidase into the phase in order to remove oxygen, for example in the form of a surface layer on a food within a gas-proof packing material. Dr. Sven Engström in our department has found that this enzyme keeps its activity during months when it is incorporated into this phase.

Literature Cited

1. Engström, L. J. Disp. Sci. Techn., in press.
2. Larsson, K. J. Phys. Chem. 1989, 73, 7304-7314.
3. Lindström, M.; Ljusberg-Wahren, H.; Larsson, K.; Borgström, B. Lipids 1981, 16, 749-754.
4. Larsson, K. J. Colloid Interface Sci. 1979, 72, 152-153.
5. Pilman, E.; Tornberg, E.; Larsson, K. J. Disp. Sci. Techn. 1982, 3, 335-349.
6. Moberger, L.; Larsson, K.; Buchheim, W.; Timmen, H. J. Disp. Sci. Techn. 1987, 8, 207-215.

RECEIVED August 7, 1990

Chapter 5

Preparation of 2-Monoglycerides

Adam W. Mazur, George D. Hiler, II, and Magda El-Nokaly

Food and Beverage Technology Division, Miami Valley Laboratories,
The Procter and Gamble Company, P.O. Box 398707,
Cincinnati, OH 45239-8707

We present a method for the selective hydrolysis of
triglycerides resulting in an efficient preparation of
2-monoacyl glycerols or specific mixtures of the latter
compounds with 1,2-diacyl glycerols. These derivatives
are important substrates for the synthesis of tailored
triglycerides. The hydrolysis reactions are carried out
in a simple biphasic system consisting of hexane, alkyl
alcohol, and enzyme-containing buffer, without added
emulsifiers. The analysis of reaction kinetics shows
that selectivity of this hydrolysis depends upon the
specific composition of solvents and the key factor in
the selectivity may be modification of the interface
properties by solvents.

The work on selective hydrolysis of triglycerides to 2-acyl
glycerols, described in this chapter, has been part of our
research on the preparation of tailored triglycerides. These are
important materials for the investigations of functional and
biological properties of pure fat components. Unfortunately,
there are no published procedures suitable for preparation on a
large scale of specific triglycerides having different three acyl
groups.
We found that the reactions between 2-monoacyl glycerols and
fatty acid anhydrides catalyzed by lipase in organic solvents can
produce diglycerides selectively with very little formation of
triglycerides (Figure 1). Acylation of diglycerides with another
acyl anhydride in the presence of an enzyme or a chemical
catalyst, such as 4-N,N-dimethylaminopyridine, produces the
required specific triglycerides. These products are also expected
to have an excess of one enantiomer since the NMR spectra of
diglycerides measured in the presence of a chiral shift reagent
indicate that the reaction in Figure 1 is stereoselective (Mazur,
A. W.; Hiler, G. D., The Procter & Gamble Co., unpublished data).

0097-6156/91/0448-0051$06.00/0
© 1991 American Chemical Society

However, the key substrates, 2-acyl glycerols, were not
readily available in multigram quantities. Typical chemical
methods (1) used for their preparation require several steps and
yields are frequently affected by isomerization of the acyl group
between 1(3)- and 2- positions of glycerol.

Another possible source of 2-monoacyl glycerols is hydrolysis
of triglycerides with 1,3-specific lipases. The reaction produces
mono- and diglycerides as intermediates (2) before finally giving
glycerol (Figure 2). Unfortunately, selectivity in a simple
aqueous hydrolysis is not easy to control and preparation of
2-monoglycerides is not practical by this route. Recently, it was
reported that the formation of microemulsion phase, by addition

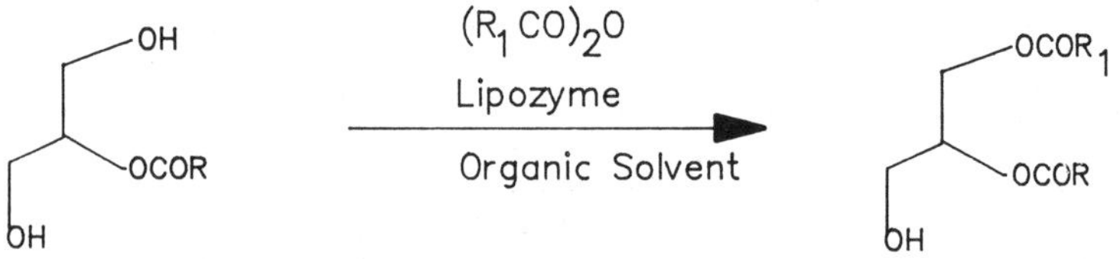

Figure 1. Regioselective Esterification of 2-Acyl Glycerols

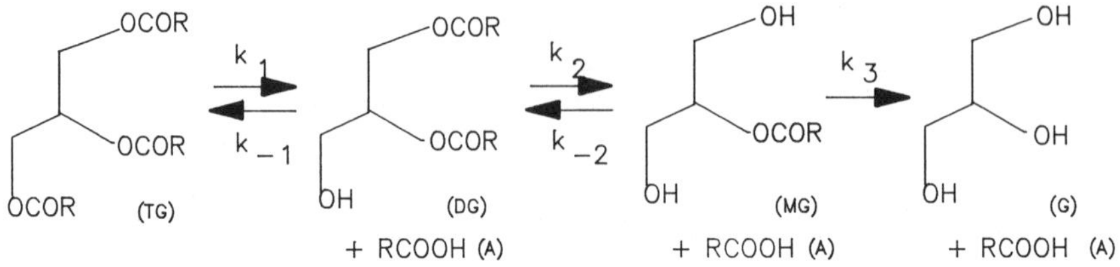

$$d[TG]/dt = -k_1[TG] + k_{-1}[DG][A]$$

$$d[DG]/dt = k_1[TG] - k_{-1}[DG][A] + k_{-2}[MG][A] - k_2[DG]$$

$$d[MG]/dt = k_2[DG] - k_{-2}[MG][A] - k_3[MG]$$

$$d[G]/dt = k_3[MG]$$

Figure 2. Hydrolysis of Triglycerides

of emulsifiers, can induce selectivity in selectivity in lipolysis and good yields of 2-monoacyl glycerols were obtained (3-4). However, this method is not convenient as the product has to be purified from the mixtures containing substantial quantities of emulsifiers.

Due to these limitations, our objective was to find an efficient method for the preparation of 2-monoacyl glycerols in large quantities.

It is known that enzymatic lipolysis takes place at the oil/water interface (5) for typical biphasic mixtures and microemulsions. The latter systems, of fine dispersion of micelles are optically transparent. Physically, they are still biphasic with the enzymes located in the micellar water pools (6), the micelle palisade, or in the interfacial film. Microemulsions can promote selectivity because association structures created by emulsifiers and solvents modify properties of the interface thus influencing enzyme activity (6-11) as well as kinetics and equilibria (7b, 12-14) of the reactions. The reactants present in the lipolysis mixtures including triglycerides, diglycerides, monoglycerides and fatty acids, can create various association structures such as liquid crystals or microemulsions depending upon a specific composition of the mixture (15-18). The presence of organic solvents is one of the important factors affecting self-association of amphiphilic molecules (19-20). We were interested in influencing the process of microstructure formation and the properties of interface by proper selection of solvent and, ultimately, controlling the selectivity of the biphasic hydrolysis without emulsifiers.

MATERIALS AND METHODS

Hexane and butanols were purchased from J.T. Baker Co., phosphate buffer from Fisher and Lipolase from Novo. The quantitative GC analysis of the reaction mixtures including triglycerides, diglycerides, monoglycerides, glycerol, and acid was done using HP 5880A instrument equipped with a capillary column (DB-1 15 m x 0.247 mm, film thickness 0.25 micron). Samples were analyzed under the following conditions: injection port temperature 340°C, split injection (60:1), flame ionization detector at 325°C. Oven temperature profile: 70°C to 300°C at 15°C/min hold 300°C for 10 min. All samples were silylated with N,O-bis-(trimethylsilyl)- trifluoroacetamide (Pierce) prior to the injection into column in order to convert glycerol and its partial esters into volatile derivatives. Changes in concentrations of these species with time were recorded and used for plotting the reaction progress curves as well as calculating the individual rates for each hydrolysis step. Quantitative analysis of the kinetic data was performed with the program GEAR Iterator (21) on the VAX 785 computer. The program was purchased from the Quantum Chemistry Exchange Programs at Indiana University. The points on the graphs are the experimental data while the continuous lines represent the computer generated fit (vide supra).

Hydrolysis of Trioctanoin in Biphasic Mixtures. Quantities of
reagents and solvents used in the reactions are shown in Table 1.
In the reactions with primary alcohol, 20 ml of 1-butanol were
used. The mixtures were stirred vigorously at 25°C. We did not
follow the pH during the reactions but GC analysis showed that at
least 85% of free acids formed in each hydrolysis partitioned
into the organic phase. After stirring was stopped, all the
reaction mixtures gave good phase separation upon standing and
only two phases, upper organic and lower aqueous, were seen. For
isolation of 2-octanoyl glycerol, organic phase was evaporated at
the temperature below 30°C to a viscous residue which was
dissolved in hexane. The product precipitating from the solution
at -78°C was filtered at this temperature and dried under
vacuum. The yield was 70%.

Hydrolysis of Other Triglycerides. Fully saturated
triglycerides, up to trimyristin ($C_{14:0}$), can be hydrolyzed
analogously to trioctanoin. Beginning with tripalmitin ($C_{16:0}$),
preparation of 2-monoglycerides according to our procedure is not
practical since hydrolysis occurs very slowly. However,
unsaturated triglycerides regardless of the chain length undergo
hydrolysis similarly to trioctanoate.

ANALYSIS OF THE REACTION KINETICS

To investigate the phenomena responsible for the selectivity of
lipolysis, we have analyzed the reaction progress curves (Figures
3 and 4 are examples) and variations in the rate constants (Table
I) as calculated from the kinetic model (Figure 2).

Effect of a Primary Alcohol. Figure 3 shows the reaction
progress curve for the hydrolysis of triglyceride in buffer
without addition of organic solvents. Tricaprylin is not soluble
in the buffer and the reaction mixture is biphasic. The reaction
reaches an equilibrium or a steady state characterized by a
relatively low degree of triglyceride hydrolysis and low
monoglyceride concentration. This confirms our earlier assumption
that little selectivity in formation of 2-monoglycerides can be
expected in the unmodified hydrolysis system. In contrast,
hydrolysis in the presence of hexane and 1-butanol (Figure 4)
produces excellent yields of 2-monoglycerides. Caprylic acid is
esterified in the latter process to butyl ester thus moving the
reaction equilibrium towards products. One could certainly argue
at this point whether the properties of interface are really
important for the displayed selectivity, or the good yield of
2-monoglyceride is solely the result of the equilibrium shift
caused by acid esterification with butanol. That the interface
modification is indeed important for the selectivity, can be
concluded from the results discussed below.

Effects of a Secondary and Longer Chain Alcohols. Upon changing
the alcohol to 2-butanol, which is not esterified in the
reaction, the selectivity of 2-monoglyceride formation was still

Table 1. Biphasic Hydrolysis of Trioctanoin
Effects of Solvent Composition, and Triglyceride Concentration

Entry	Solvent	TG	k_1	k_{-1}	k_2	k_{-2}	k_3
		[M]			10^{-3}		
1		None	2.03	332.0	4.69	72.9	7.6	0.77
2	Hexane		0.70	225.0	20.30	108.0	44.5	13.70
3		2-Butanol	1.40	346.0	2.20	106.0	4.7	<0.01
4	Hexane	2-Butanol	0.72	217.0	5.59	78.7	10.9	2.14
5	Hexane	2-Butanol	0.60	214.0	5.95	89.6	12.9	<0.01
6	Hexane	2-Butanol	0.41	256.0	7.22	78.0	11.4	<0.01
7	Hexane	2-Butanol	0.31	204.0	7.41	94.1	15.3	0.43
8	Hexane	2-Butanol	0.21	202.0	8.61	70.3	11.3	1.73

Volumes of Solvents: Hexane 60 ml, 2- Butanol 20 ml, Phosphate
                     Buffer 50 mM, pH 7.0 20ml
Amount of lipolase: 31.8 x 10^3 lipase units

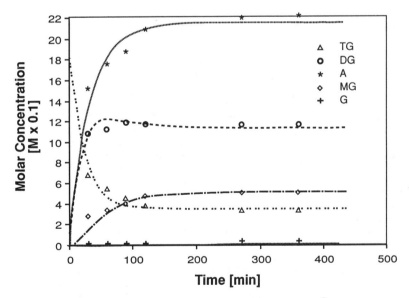

Figure 3. Hydrolysis of Tricaprylin, 25°C
Phosphate Buffer 50 mM, pH 7.0

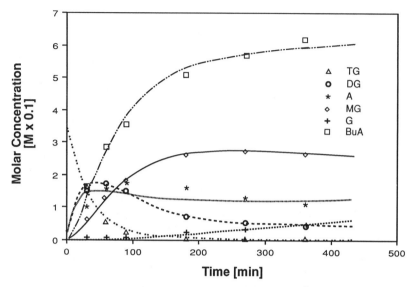

Figure 4. Hydrolysis of Tricaprylin, 25°C
Hexane, 1-Butanol, Phosphate Buffer 50 mM, pH 7.0

very good, if low initial concentrations (3 - 5 %) of
triglyceride were used. Figures 5 and 6 show how the initial
concentration of triglycerides effects the final composition of
the reaction mixture. Moreover, less glycerol was produced in
the presence of 2-butanol compared to the reactions with
1-butanol. Interestingly, these results are similar to those
reported for the reactions in microemulsions (4). On the other
hand, if the solvent was hexane but without alcohol the
selectivity was very poor regardless of the initial substrate
concentration. The latter hydrolysis produced glycerol much more
rapidly and gave less 2-monoglycerides than the process shown on
Figure 3. Further indication of the importance of the interfacial
properties is that butanols and pentanols have been highly
effective in inducing selectivity while higher alcohols
drastically reduced the rate of hydrolysis. Medium chain alcohols
are known to be particularly effective in modifying such
interfacial properties as surface tension, fluidity and curvature
(22).

Kinetic Model. Following the qualitative analysis of the
reaction progress curves, an attempt to quantify the effects of
solvents composition on the selectivity of hydrolysis was made.
The kinetic model shown in Figure 2 was analyzed. This model is
highly simplified, the rate constants k are in fact functions of
a number of parameters. Lipase is one of the reactants (catalyst)
in each step and all k's depend upon the enzyme concentration
(activity). Likewise, the forward hydrolysis rates, k_1, k_2,
and k_3 depend upon the water concentration. These two

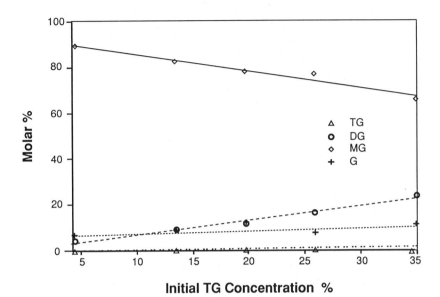

Figure 5. Hydrolysis of Tricaprylin, 25°C
Hexane, 1-Butanol, Phosphate Buffer 50 mM, pH 7.0

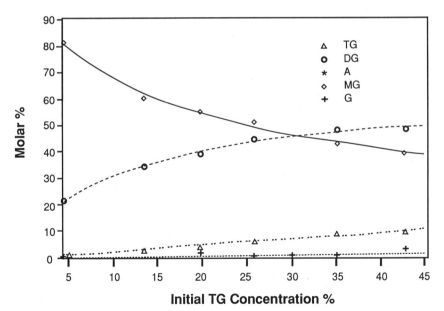

Figure 6. Hydrolysis of Tricaprylin, 25°C
Hexane, 2-Butanol, Phosphate Buffer 50 mM, pH 7.0

parameters remain constant in most of the discussed reactions.
The rate of glycerol formation, k_3, is a function of factors
responsible for two processes: isomerization of 2-monoacyl
glycerol to 1-monoacyl glycerol and hydrolysis of the latter to
glycerol. This initial isomerization is necessary because
1,3-specific lipase can not cleave the acyl group from the
2-position in glycerol. The lipase specificity also explains
irreversibility of the last step; this enzyme would not esterify
glycerol to 2-monoglyceride but only to 1-mono, or
1,3-diglycerides. Finally, our rate constants can be influenced
by the association structures formed in the reaction mixture.
Therefore, variations of the individual rates upon changing the
reaction conditions are expected to give us some insight into the
sources of possible selectivity in lipolysis.

Quantitative Analysis of the Kinetic Model. Calculations of the
rate constants for the described model were done using the
computer program "GEAR Iterator" (21). The program numerically
integrates the reaction rate equations using an arbitrary initial
set of k's and iteratively fits equations to the experimental
data. In effect, sets of rate constants with their standard
deviations are calculated for each graph. In most cases, our
calculations converged on the values of rate constants having
small standard deviations. Exceptions were the reactions in the
presence of primary alcohols where the iterator failed to find a
satisfactory fit between the kinetic model and the experimental
results. In these instances, we tried to improve the kinetic
model by adding a term describing a reversible esterification of
1-butanol but no improvement in the quality of fit was achieved.
Consequently, we were not able to analyze quantitatively the
reactions with primary alcohols.
    Table I shows the values of the rate constants calculated by
the GEAR Iterator for selected reactions. It is interesting that
the hydrolysis rates, k_1 and k_2, do not change significantly
with conditions. In contrary, the re-esterification rate
constants k_{-1}, k_{-2} and, in particular, the glycerol formation
rate constant k_3 change substantially. Hexane (entry 2)
apparently facilitates the re-esterification and glycerol
formation reactions but the presence of 2-butanol (entry 4)
reduces the effects of hexane. The influence of the initial
triglyceride concentration on k_{-1} is shown on Figure 7. Since
an increase of the initial triglyceride concentration is
equivalent to an increase in acid concentration during
hydrolysis, the variation of this re-esterification rate can be
interpreted as a decrease in acid activity or availability for
the esterification when the acid concentration increases. These
observation indicates that the acid may create association
structures such as aggregates which change the properties of the
acid, e.g. by increasing a degree of ionization, thus making this
substrate less active in the esterification reaction.
    It should be pointed out the mathematical methods presently
applied for solving sets of nonlinear differential equations,
such as used in chemical kinetics have limitations. In general,
these equations do not have analytical solutions and there is no
mathematical method to prove that the approximate solution is
unique. Therefore, the results of calculations should be treated

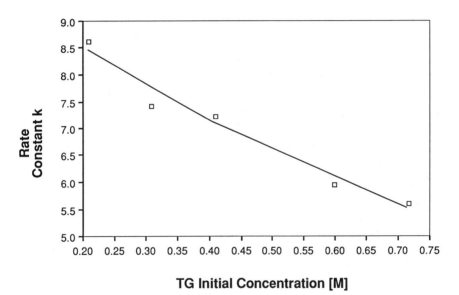

TG Initial Concentration [M]

Figure 7.   Rate Constant k_{-1} vs. TG Initial Concentration

with caution and mechanistic interpretations based on them should
be verified by other methods. Accordingly, our results showing
selectivity of the reaction and consequently yields of
2-monoglycerides might depend upon the kind of organic solvent
were further confirmed experimentally.

The discussed results indicate that a character of interface
is the key parameter for the selectivity in triglyceride
hydrolysis. Monoglycerides as well as acids create possibly
association structures at the interface under certain conditions,
limiting their activity or availability for the reactions.
Another possibility is the interface composition has an effect on
the enzyme activity e.g. by modifying the Michaelis constants for
each hydrolysis step.

SUMMARY AND CONCLUSIONS

Highly selective hydrolysis of triglycerides giving good yields
of 2-acyl glycerols can be carried out in a simple biphasic
system consisting of hexane, aliphatic alcohol, and an aqueous
buffer containing 1,3-specific lipase. Our results indicate that
selectivity of this hydrolysis depends upon interfacial
properties which can be modified by organic solvents. Lower
aliphatic alcohols with exception of propanols effectively induce
selectivity while longer chain alcohols, begining with hexanols,
drastically reduce the rate of hydrolysis. Medium chain alcohols
are known very good promoters of microemulsion formation, being
particularly effective in modifying such interfacial properties
as surface tension, fluidity and curvature. Analysis of the rate
constants shows that the selectivity is controlled by changes in
the magnitude of re-esterification rates as well as the rate of
glycerol formation.

LITERATURE CITED

1. Bhati, A.; Hamilton, R. J.; Steven, D. A. in Fats Oil: Chem.
    Technol.; Hamilton, R. J. and Bhati, A., Eds.; Appl. Sci.:
    London, 1980; Chapter 4, p 59 - 70.
2. Brockerhoff, H.; Jensen, R. G. in Lipolytic Enzymes,
    Academic, 1974.
3. Holmberg, K.; Osterberg, E. J. Am. Oil Chem. Soc., 1988,
    65(9), 1544 - 1548.
4. Osterberg, E.; Ristoff, C.; Holmberg, K. Tenside Surfactants
    Deter., 1988, 25(5), 293 - 297.
5. Lipases, Borgstrom, B.; Brockman, H. L., Ed., Elsevier, 1984.
6. Luisi, P. L.; Magid, L. J. C.R.C. Crit. Rev. Biochem., 1986,
    20(4), 409 - 474.
7a. Luisi, P. L.; Steinmann-Hofmann, B. Methods Enzymol., 1987,
    136, 188 - 216.
7b. Laane, C.; Hilhorst, R.; and Veeger, C. Methods Enzymol.,
    1987, 136, 216 - 229.
8. Castro, M. J. M.; Cabral, J. M. S. Biotechnol. Adv., 1988, 6,
    151 - 167.

9. Kabanov, A. V.; et. al. J. Theor. Biol., 1988, 133, 327 - 343.
10. Martinek, K.; et. al. Eur. J. Biochem., 1986, 15, 453 - 468.
11. Schomaker, R.; Robinson, B. H.; Fletcher, P. D. J. J. Chem. Soc., Faraday Trans. 1, 1988, 84(12), 4203 - 4212.
12. Eggers, D. K.; Blanch, H. W.; Prausnitz, J. M. Enzyme Microb. Technol., 1989, 11, 84 - 89.
13. Martinek, K.; Semenov, A. N.; Berezin, J. V. Biochim. Biophys. Acta, 1981, 658, 76 - 89.
14. Martinek, K.; Semenov, A. N. Biochim. Biophys. Acta, 1981, 658, 90 - 101.
15. Larsson, K. Z. Physical Chem., 1967, 56, 173 - 198.
16. Ljusberg-Wahren, H.; Herslof, M.; and Larsson, K. Chem. Phys. Lipids, 1983, 33, 211 - 214.
17. Xi-Kui, J. Acc. Chem. Res., 1988, 21, 362 - 367.
18. Fedorov, A. B.; Zaichenko, L. P.; Abramzon, A. A.; Proskuryakov, V. A. Zhur. Prikl. Khim., 1984, 5, 1093 - 1097.
19. Kunieda, H. J. Colloid Interface Sci., 1989, 133(1), 237 - 243.
20. Kunieda, H.; Asaoka, H.; Shinoda, K. J. Phys. Chem., 1988, 92, 185 - 189.
21. McKinney, R. J.; Weigert, F. J. Gear: Gear Algoritm Integration of Chemical Kinetics Equations (QCMPO 22), Quantum Chemistry Program Exchange, Indiana University.
22. Bansal, V. K.; Shah, D. O.; O'Connel, J. P., J. Colloid Interface Sci., 1980, 75(2), 462 - 475.

RECEIVED September 9, 1990

Chapter 6

Dioctyl Sodium Sulfosuccinate–Sorbitan Monolaurate Microemulsions

D. W. Osborne, C. V. Pesheck, and R. J. Chipman

Drug Delivery Research and Development, The Upjohn Company, Kalamazoo, MI 49001

Ternary and pseudo-ternary phase behavior studies of the nonionic surfactant sorbitan monolaurate (Arlacel 20), the ionic surfactant dioctyl sodium sulfosuccinate (AOT), hexadecane, and water have been completed. Both the waxy semisolid AOT and the commercially available liquid form of this surfactant AOT-75 (75% AOT, 20% water, and 5% ethanol) were investigated. Extensive microemulsion regions capable of incorporating more than 55% water were found for the optimum surfactant mix between 6/4 and 7/3 by weight AOT-75/Arlacel 20, while only insignificant water incorporation was seen for either surfactant alone. Rheological properties of the surfactant mixture were also determined to aid in evaluation of the ultimate manufacturability of the microemulsion. Surfactant mixtures containing less than 60 weight percent of the more viscous surfactant, i.e., sorbitan laurate, showed insignificant differences in viscosity and could readily be mixed for large-scale manufacture.

The use of surfactants to aid in drug solubilization and delivery has been successful for numerous routes of drug administration (1,2). Most recently, microemulsion systems have been considered for use as oral, injectable and topical dosage forms (3). While the microemulsion system studied here was initially intended for use as a topical drug delivery vehicle, the AOT-Arlacel 20 surfactant combination may be of interest for the researcher formulating food microemulsions.

The advantages of a microemulsion over emulsions, suspensions, or solutions is either improved stability or solubilization characteristics. Since microemulsions are in most cases thermodynamically stable, separation problems that frequently occur with emulsions are avoided. Microemulsions also have the potential ability to solubilize both lipophilic and hydrophilic species, which allows for a variety of flavoring and coloring agents having vastly different physical properties to be dissolved/solubilized within a system. A major disadvantage of

0097–6156/91/0448–0062$06.00/0
© 1991 American Chemical Society

microemulsion systems for ingestion is the traditional need for a medium chain alcohol such as pentanol to function as a cosurfactant (4). Since alcohols of this chain length tend to posses unacceptable toxicity/irritation profiles (Table I), their use in foods is very limited. Additionally the taste associated with these materials is unacceptable. While the palatability of any microemulsion containing AOT may be unacceptable, the determination of the phase behavior provides the framework for further studies specific to food science.

The results of Johnson and Shah (5) suggest that the combination of dioctyl sodium sulfosuccinate (AOT) and sorbitan monolaurate results in a microemulsion capable of obtaining a molar ratio of water to surfactant of approximately 50. AOT and sorbitan monostearate have food additive status in the United States. While AOT:sorbitan monolaurate mixtures are the focus of this study, similar trends in the phase behavior are expected for AOT:sorbitan monostearate mixtures. The phase behavior of the surfactant AOT has been extensively studied (6-8). By eliminating the need for a medium chain alcohol such as pentanol, this microemulsion system avoids use of the biologically aggresive medium chain alcohols.

Initial phase behavior studies of the AOT-sorbitan laurate microemulsions indicated a severe practical limitation of the system. Long mixing times were required for the waxy semi-solid surfactant AOT to mix with the viscous liquid Arlacel 20. In attempts to avoid the slow mixing of the surfactants, the commercially available liquid surfactant AOT-75 was used instead of AOT. AOT-75 is a solution containing 75% AOT, 20% water and 5% ethanol. Unlike the medium chain alcohols, ethanol ingestion is of known consequence. Thus in these small amounts, use of ethanol is completely acceptable. By determining the rheological properties of the miscible AOT-75/Arlacel 20 mixture, and characterizing the phase behavior of this surfactant mixture when combined with hexadecane and water, optimal compositions for a low-alcohol microemulsion were determined.

Table I. Safety Considerations of Medium Chain Length Alcohols

Alcohol	Oral LD50 in Rat (mg/kg)	Eye Irritation	Skin Irritation
2-Butanol (least irritating of 4 isomers)	6480	Moderate	Threshold Conc.7.8%
Hexanol	720	Severe	Mild
Octanol	1790*	Moderate	Threshold Conc.12%
Decanol	472	Severe	Severe

*Oral LD50 for mouse, oral LD50 for humans 0.5-5 gm/kg

Materials and Methods

The AOT and sorbitan monolaurate were obtained commercially as
Docusate Sodium USP (American Cyanamid, Bridgewater, NJ) and
Arlacel 20 (ICI Americas Inc., Wilmington, DE). Dioctyl sodium
sulfosuccinate is the USP grade of the substance sodium 1,4-bis(2-
ethylhexyl)-sulfosuccinate, and is also marketed as Aerosol-OT.
Aerosol OT-75 (American Cyanamid, Wayne, NJ) is a liquid form of
dioctyl sodium sulfosuccinate that is 75% surfactant, 20% water
and 5% ethanol. Hexadecane (Certified 99.8 mole %) was obtained
from Fisher. These materials were all used as received. USP
purified water was further purified using a MILLI-Q Reagent water
System until a resistance greater than 15 megohms-cm was
maintained.

Phase boundaries of the liquid isotropic phase of the AOT
systems were determined by mixing two or three of the components
and titrating with the remaining component, a method that also
gave approximate information about the liquid crystalline regions.
The precise location of the phase boundaries were obtained by
thoroughly mixing selected compositions near the titrated phase
boundaries and allowing them to equilibrate for more than a week.
For the more viscous phases, thorough mixing was achieved by
repeatedly centrifuging the matrix through a constriction in a
sealed 7 mm glass tube.

Phase behavior for the AOT-75 microemulsion system was
facilitated by use of a Perkin-Elmer Laboratory Robotics System
(9) with an added EFD dispenser (EFD, Inc., 977 Waterman Avenue,
East Providence, RI 02914) capable of dispensing viscous
materials. This system calculated and dispensed the correct
volume of water and hexadecane, weighing each after dispensing,
and then added the surfactant mixture to the correct weight.
Original samples were created at every five weight percent
increments of each component over the area of interest.
Additional samples were mixed as needed to clarify phase
boundaries.

Once mixed and equilibrated, the samples were examined
visually and between crossed polars at controlled room
temperatures of 24 ± 2°C. Certain samples near the enclosed two-
phase region were quite temperature sensitive; these were
equilibrated in a temperature controlled chamber at 25 ± 0.2°C.
Any cloudiness or birefringence was noted. Observations were
repeated until no further changes were noted. Since only the one-
phase, microemulsion region was of interest, no attempt was made
to determine the number of phases in the cloudy samples or their
structure. The position of the boundaries have been determined
within 2%. Viscosity data was obtained on a Contraves Rheomat 108
using spindles #1, 2, 3 and shear rates of 17.6 to 355 sec-1 at
constant ambient temperature.

Results

Figure 1 provides the base diagram for the Water-Arlacel 20-AOT
system. The most striking feature of this diagram is that less
than 5% (wt/wt) of Arlacel 20 can be solubilized in the water/AOT

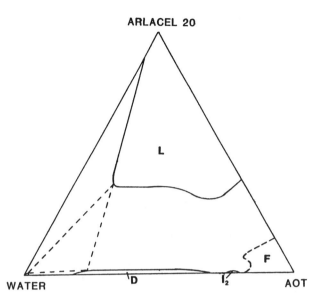

ARLACEL 20

L

F

WATER   D   I₂   AOT

Figure 1. Ternary phase behavior for the Water-Arlacel 20-AOT system at 25°C. The single phase regions are denoted by L for a clear, colorless, isotropic liquid, D for the lamellar liquid crystalline phase, I₂ for the cubic liquid crystalline phase, and F for the inverse hexagonal liquid crystalline phase.

lamellar structure (region D) at equilibrium. For the water/AOT weight ratios from 4/1 to 3/7, addition of 4-8% Arlacel 20 resulted in separation of an isotropic phase which could be detected microscopically after approximately one week. The addition of larger amounts of Arlacel 20 had a more immediate destabilizing effect.

The maximum water solubilization obtainable for three Arlacel 20/AOT ratios in hexadecane is shown in figure 2. The 73/27 Arlacel 20/AOT surfactant ratio that gave maximum water solubilization for the base diagram (figure 1) did not give maximum water solubilization after addition of more than 20% hexadecane. Similarly, the 85/15 Arlacel 20/AOT surfactant ratio had limited ability to solubilize water. However, increasing the amount of AOT to a 60/40 Arlacel 20/AOT surfactant ratio provided a large isotropic region capable of solubilizing better than 60% water. More importantly, this system was able to solubilize significantly more water at higher hexadecane concentrations than the systems with larger Arlacel 20 surfactant ratios.

Further rheology and phase behavior studies used the commercially available liquid form of the waxy solid anionic surfactant consisting of 75% AOT, 20% water and 5% ethanol

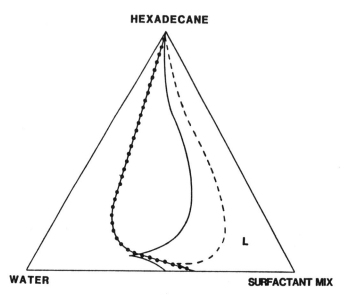

Figure 2.   Pseudo-ternary phase behavior for the Water-
Hexadecane-Arlacel 20-AOT system, where the Arlacel 20:AOT
weight ratio is 60:40 (-●─●-); 72.5:27.5 (────); and 85:15
(- - -). The single phase clear, isotropic microemulsion side
of the diagram is labeled by L.

mixture, i.e. AOT-75. Figure 3 shows the viscosity of the AOT-
75/Arlacel 20 mixtures measured at an apparent shear rate of 64.6
s^{-1}. As seen, increases in the amount of Arlacel 20 contained in
the surfactant mixture only slightly increases the viscosity until
60 to 70% Arlacel 20, i.e., sorbitan laurate, is added to AOT-75.
At the highest Arlacel 20 concentrations, the viscosity increases
significantly.   Figure 4 shows the variation of viscosity with
changes of the shear rate.   Shear thinning was detected for each
of the surfactant mixtures, with the higher viscosity samples
being thinned significantly by relatively low shear rates.
The extent of the microemulsion region between 100% AOT 75
and 60:40 AOT-75:Arlacel 20 is shown in figure 5. This one-phase
region grows in area smoothly.   All surfactant ratios except the
100% AOT-75 are completely miscible with hexadecane, and the
amount of water soluble in the surfactant-hexadecane solutions
grows as the amount of Arlacel 20 increases.

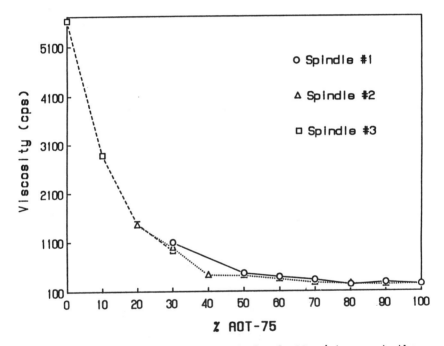

Figure 3.   Viscosity of AOT-75/Arlacel 20 mixtures at the constant apparent shear rate of 64.6 sec^{-1}.

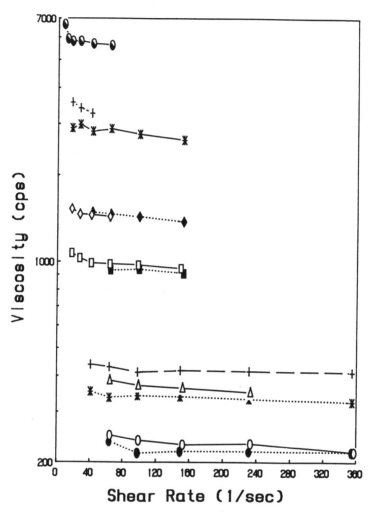

Figure 4.   Shear Rate (Apparent) Dependent Viscosity for AOT-
75/Arlacel 20 Mixtures:

O—O	100% AOT-75 Spindle 1
●—●	100% AOT-75 Spindle 2
△—△	60% AOT-75/40% Arlacel 20 Spindle 1
⋯	60% AOT-75/40% Arlacel 20 Spindle 2
+—+	40% AOT-75/60% Arlacel 20 Spindle 2
□—□	30% AOT-75/70% Arlacel 20 Spindle 2
■⋯■	30% AOT-75/70% Arlacel 20 Spindle 3
◇—◇	20% AOT-75/80% Arlacel 20 Spindle 2
◆⋯◆	20% AOT-75/80% Arlacel 20 Spindle 3
+⋯+	10% AOT-75/90% Arlacel 20 Spindle 2
—	10% AOT-75/90% Arlacel 20 Spindle 3
O—O	100% Arlacel 20 Spindle 3

HEXADECANE

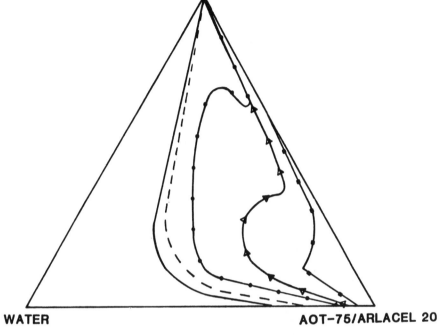

WATER **AOT-75/ARLACEL 20**

Figure 5. Pseudo-ternary phase behavior for the water/hexadecane/ AOT-75/Arlacel 20 System at 24° ± 2°C. The ratios of AOT-75 to Arlacel 20 (sorbitan laurate) are as follows: —■—■— 100% AOT-75; —◄—◄— 90/10 AOT-75/Arlacel 20; —●—●— 80/20 AOT-75/Arlacel 20; - - - 70/30 AOT-75/Arlacel 20; ———— 60/40 AOT-65/Arlacel 20.

Figures 6 and 7 show the transition region near the maximum water solubility microemulsion. The maximum water solubility appears to occur between 55/45 and 60/40 AOT-75/Arlacel 20 ratios and is approximately the same for hexadecane to surfactant mixture ratios between 20/80 and 50/50. At 54% AOT-75, a two-phase region appears inside the one-phase microemulsion region. The samples in this two-phase region, and nearby in the one-phase region, were temperature sensitive and were examined in a temperature controlled air chamber. As the fraction of AOT-75 decreases further, the microemulsion region also decreases in size (figure 8). The cusp that appears implies a three-phase region adjacent to it; evidence of this region was observed in several samples, but not examined closely.

Hexadecane was replaced with mineral oil in order to investigate a more applied system. As seen in figure 9, the progression of the phase behavior for the mineral oil system is analogous to the hexadecane system. At a near optimal AOT-

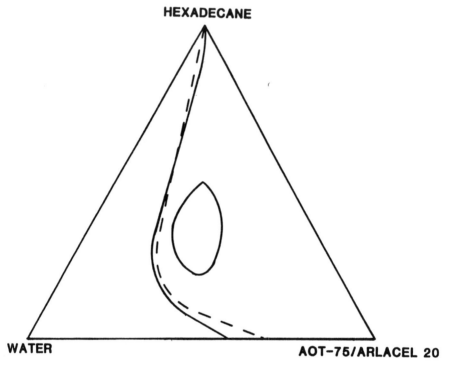

HEXADECANE

WATER   AOT-75/ARLACEL 20

Figure 6.   Pseudo-ternary   phase   diagram   for   the
water/hexadecane/ AOT-75/Arlacel 20 system at 25 $\pm$ 0.2°C.
Dashed line is for 57/43 AOT-75/Arlacel 20 ratio, while the
solid line is for the 54/46 AOT-75/Arlacel 20 ratio. Note the
two phase region within the single phase region for the 54/46
AOT-75/Arlacel 20 system.

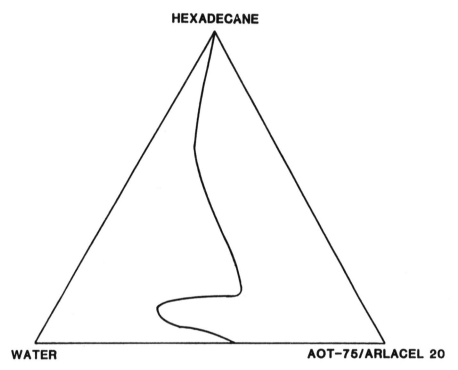

Figure 7. Pseudo-ternary phase diagram for the water/hexadecane/ AOT-75/Arlacel 20 system at 24 ± 2°C for the 50/50 AOT-75/Arlacel 20 mixture.

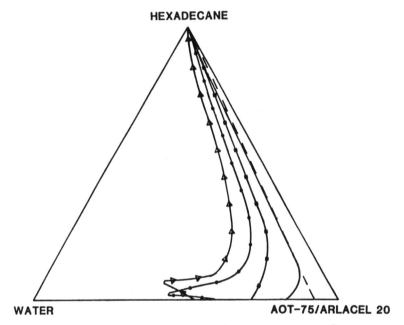

Figure 8.   Pseudo-ternary phase diagram at 24 ± 2°C for the
water/hexadecane/AOT-75/Arlacel 20 system.   The ratios of AOT-
75 to Arlacel 20 are as follows:▷—▷—40/60 AOT-75/Arlacel 20;
—●—●—30/70 AOT-75/Arlacel 20; —■—■—20/80 AOT-75/Arlacel 20;
————10/90 AOT-75/Arlacel 20; and - - - 100% Arlacel 20.

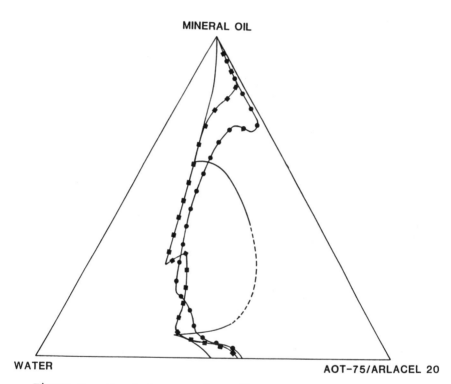

Figure 9.   Pseudo-ternary phase diagram for the water/mineral oil/AOT-75/Arlacel 20 system at 25 ± 2°C. The weight ratios of AOT-75 to Arlacel 20 are as follows:—●—●—50/50;—■—■—45/55 AOT-75/Arlacel 20;———40/60 AOT-75/Arlacel 20.

75/Arlacel 20 ratio the microemulsion region has a broad-rounded phase boundary with maximum water incorporation of approximately 55 weight percent. As the amount of AOT is decreased from this near optimal ratio, the single phase microemulsion region is dramatically interrupted. While the progression of phase behavior changes apparently differs insignificantly when hexadecane is replaced with mineral oil, the near optimal surfactant ratio is shifted from 60/40 AOT-75/Arlacel 20 for hexadecane to 40/60 AOT-75/Arlacel 20 for mineral oil.

Discussion

The base diagram as shown in figure 1 provides insight concerning the surfactant interactions that allow the formation of an alcohol-free microemulsion. The addition of Arlacel 20, i.e., sorbitan monolaurate, destabilizes the structure of the AOT/water lamellar phase. This disordering effect decreases the lamellar region while complementarily increasing the size of the isotropic single phase region. Thus, for the water, Arlacel 20, AOT base diagram, it appears that Arlacel 20 functions as a cosurfactant similar in mechanism to that of medium chain alcohols in anionic soap systems. Addition of the cosurfactant destabilizes the bilayer structured lamellar liquid crystalline phase, resulting in an increase in the compositional range over which the microemulsion is present. In contrast to the anionic soap systems (10), medium chain alcohols do not function as cosurfactants when added to the aqueous AOT system. As seen in figures 10 through 12 (11), the size of the lamellar liquid crystal phase is not reduced to the degree seen upon addition of Arlacel 20. While the microemulsion region is large when either hexanol or octanol is added as the third component, the most dominant features of these diagrams are the large cubic and inverse hexagonal liquid crystalline regions. The stability of these stiff phases shape the boundaries of the microemulsion region. The water/butanol/AOT diagram shown in figure 10 is characterized by a continuous solubility region between the butanol and water corners. This phase behavior is typical for butanol systems and the resulting fluid bicontinuous phase may be less likely to have the desired ability to solubilize both lipophilic and hydrophilic additives.

Addition of hexadecane to the AOT, Arlacel 20 system provides for a systematic characterization of the alcohol-free microemulsion. As seen in figure 2, the surfactant ratio has a significant influence on the water solubilization capacity of the system. If high water solubilization over a range of mixed surfactant/oil ratios is required, then the Arlacel 20/AOT ratio should be less than 73/27, i.e., the surfactant ratio that results in the maximum water solubility in the base diagram (figure 1).

The low solubility of the waxy semisolid AOT in water and the subsequent formation of liquid crystalline phases at higher AOT composition has been well characterized in the literature (6). AOT-75 is a liquid because the addition of 5 weight percent ethanol to the 75/20 AOT/water ratio prevents the formation of the cubic liquid crystalline phase that exists from 75 to 80% AOT in

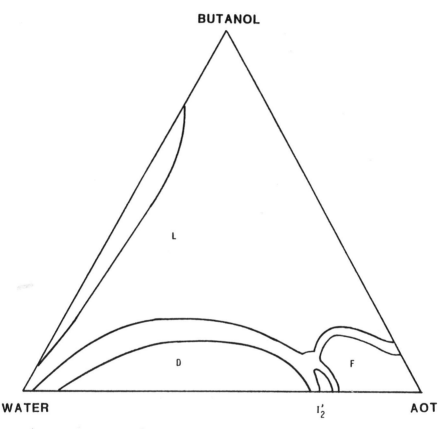

Figure 10.   Phase diagram for the system water, butanol, AOT at 24 ± 2°C.   L is the continuous solution region, D is the lamellar liquid crystal, I_2 is the cubic liquid crystal, and F is the reversed hexagonal liquid crystal.

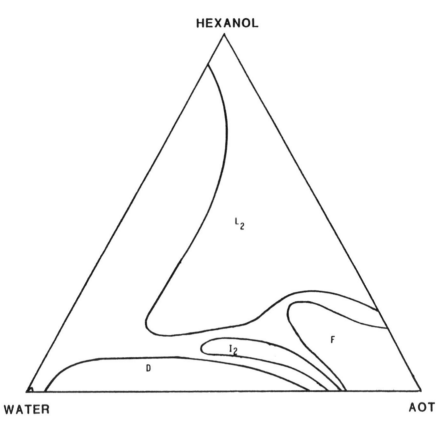

Figure 11.   Phase diagram for the system water, hexanol, AOT at
24 ± 2°C.   L$_2$ is the inverse micellar (microemulsion) region.
Other phases are labeled the same as in Figure 10.

water. Thus, AOT-75 remains a liquid (viscosity approximately 200
cps) provided ethanol is not lost through evaporation.
    As seen in figure 3, addition of Arlacel 20 (viscosity
approximately 5700 cps) does not dramatically increase the
viscosity of AOT-75 until less than 40% AOT-75 is present, i.e.,
greater than 60% Arlacel 20.   Thus, the mixing problems
characteristic of the waxy 100% AOT are not encountered with AOT-
75. Furthermore, the rheology studies indicate that the most
readily combined binary surfactant mixtures will contain greater
than 40% AOT-75.   Since microemulsions are thermodynamically
stable, and their formulation is reversible with regard to
temperature, mixing of 100% AOT could be facilitated by heating.
However, studies indicating surfactant hydrolysis upon long term
storage discourages this approach (12).
    The phase behavior results plotted in Figure 5-8 exhibit
increases in the amount of water incorporated for the system that

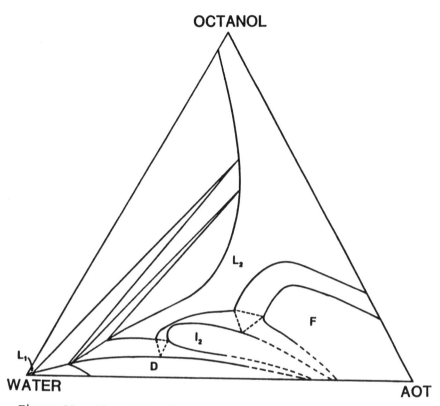

OCTANOL

L₂

F

I₂

L₁

D

WATER

AOT

Figure 12. Phase behavior for the water, octanol, AOT system at 24 ± 2°C. Phases are labeled as in Figure 10.

contains greater than 40% AOT-75. Comparison of figures 5-8 with figure 2 shows that replacing the waxy semisolid AOT with the liquid AOT-75 does not alter the trends in the phase behavior. The small amount of ethanol present in the microemulsion appears to have minimal effect on the microemulsion. Although the narrow extensions found in Figure 5 can incorporate approximately 55% water, the broad-rounded boundary characteristic of 57-70% AOT-75 is much more desirable since maximum water content can be maintained over a greater range of added hydrocarbon. For AOT-75 concentrations above the 70% AOT-75/30% Arlacel 20 mixture, the amount of water solubilized decreases, once again indicating that it is the blend of these two surfactants that produce the microemulsion region capable of the greatest water incorporation. When hexadecane is replaced by mineral oil, the progression of phase behavior is not changed, but the ratio of AOT-75/Arlacel 20 is shifted to lower AOT-75 concentrations. This emphasizes the

utility of studying a pure hydrocarbon to facilitate
interpretation, while expecting that more suitable hydrophobic
oils will be used for the initial development studies.
The unique phase behavior seen for Figures 6 and 8 deserves
further comment. It is not surprising that a mixed surfactant
system would be characterized by an ideal surfactant ratio for
incorporating water into the microemulsion. As shown in Figure 7,
the two phase region found within the single phase microemulsion
suggests that critical points may be present in this region of the
quaternary plot. It is also the 54/46 ratio of AOT-75/Arlacel 20
that provides the maximum incorporation of water. However, the
extreme temperature sensitivity of the phase behavior at this
surfactant ratio makes the 50-60% AOT-75 in Arlacel 20 systems of
limited practical use. Since single phase systems are desired for
stable formulations, slightly larger amounts of AOT-75 (60-70%)
should be mixed with Arlacel 20 to formulate a microemulsion
capable of incorporating large amounts of water. By thoroughly
characterizing the phase behavior of this multiple component
system, further efforts toward the formulation of a low-alcohol
microemulsion is possible. It should be noted that such thorough
characterization of a four component system, especially one of
practical concern, is relatively rare in the open literature (13-
15). It must be noted that AOT does not require the presence of a
cosurfactant to form a microemulsion. The microemulsion region
resulting from water-hydrocarbon-AOT systems has been extensively
studied by scattering techniques (16,17). These systems do not
tend to incorporate as much water as when Arlacel 20 is used as a
cosurfactant. Furthermore, the highest water incorporation for
these three component microemulsions tends to be relatively narrow
extensions toward the water corner. Thus, the three component
microemulsions, while likely well suited for some applications,
are not expected to be as widely useful as the Arlacel 20/AOT
microemulsions.

Conclusion

Arlacel 20 functions as a cosurfactant for the AOT system by
preventing the formation of the lamellar liquid crystalline phase
at higher water concentrations. This allows large incorporation
of water at the optimal AOT/Arlacel 20 ratio. The problem
associated with mixing rates for alcohol-free microemulsions has
been resolved by using the commercially available AOT-75
surfactant solution. The greatest water incorporation with the
least possibility of phase separation can be obtained with AOT-
75/Arlacel 20 ratios between 6/4 and 7/3 by weight.

Literature Cited

1. Attwood, D.; Florence, A.T. Surfactant Systems: Their
    Chemistry, Pharmacy and Biology; Chapman and Hall: New York,
    1983.
2. Florence, A.T. Drug Pharm. Sci. 1981, 12, 15.

3. Bhargrave, N.H.; Narurkar, A.; Lieb, L.M. <u>Pharm. Tech.</u> 1987, <u>11</u>(3), 46.

4. Friberg, S.E.;Burasczenska, I. <u>Progr. Colloid & Polymer Sci.</u>, 1978, <u>63</u>, 1.

5. Johnson, K.A.; Shah, D.O. <u>J. Colloid Interface Sci.</u>, 1985, <u>107</u>, 269.

6. Frances, E.I.; Hart, T.J. <u>J. Colloid Interface Sci.</u>, 1983, <u>94</u>, 1.

7. Fontell, K. <u>J. Colloid Interface Sci.</u>, 1973, <u>44</u>, 318.

8. Osborne, D.W.; Middleton, K.A.; Rogers, R.L. <u>J. Disp. Sci. Tech.</u>, 1988, <u>9</u>(4), 415.

9. Pesheck, C.V.; O'Neill, K.J.; Osborne, D.W. <u>J. Colloid Interface Sci.</u>, 1987,<u>119</u>, 289.

10. Ekwall, P. In <u>Advances in Liquid Crystals</u>; G.H. Brown, Ed.; Academic Press, New York, 1975, Vol. 1, p 1.

11. Osborne, D.W.; Ward, A.J.I.; O'Neill, K.J. In <u>Topical Drug Delivery Formulations</u>; D.W. Osborne and A.H. Amann, Eds.; Marcel Dekker: New York, 1990; Chapter 17, p. 349.

12. Delord, P.; Larche, F.C. <u>J. Colloid Interface Sci.</u>, 1984, <u>98</u>, 277.

13. Friberg, S.E.; Venable, R.L. In <u>Encyclopedia of Emulsion Technology</u>; P. Becher, Ed.; Marcel Dekker: New York, 1983; Vol. 1, p. 287.

14. Ng, S.M.; Frank, S.G. <u>J. Disp. Sci. Tech.</u>, 1982, <u>3</u>(3), 217.

15. Baker, R.C.; Florence, A.T.; Tadros, T.F.; Wood, R.M. <u>J. Colloid Interface Sci.</u>, 1984, <u>100</u>, 311.

16. Robinson, B.H.; Toprakcioglu, C.; Dore, J.C.; Cheiux, P. <u>J. Chem. Soc., Faraday Trans. I.</u>, 1984, <u>80</u>, 13.

17. Kotharchyk, M.; Chen, S.H.; Huang, J.S.; Kim, M.W. <u>Phys. Rev. A</u>, 1984, <u>29</u>, 2054.

RECEIVED August 7, 1990

EMULSIONS IN FOODS

Chapter 7

Sucrose Esters as Emulsion Stabilizers

Thelma M. Herrington, Brian R. Midmore, and Sarabjit S. Sahi

Department of Chemistry, The University of Reading, Reading RG6 2AD, United Kingdom

Studies of the behaviour of some sucrose esters, both in the pure state and in solution in water and n-decane, have been undertaken. The sucrose esters used were sucrose mono- and di-laurate and mono- and di-oleate, all specially synthesised, and a purified commercial surfactant, sucrose monotallowate. The pure surfactants exhibited thermotropic liquid crystalline behaviour and lyotropic mesophases were formed with water and n-decane. The extent of micellar aggregation for sucrose monolaurate and monooleate was determined in aqueous solution over the temperature range 0 to 70°C by freezing point and vapour pressure methods. Model emulsion experiments on sucrose monolaurate and monotallowate, by studying the equilibrium thickness of an aqueous surfactant film between two oil droplets, showed that increasing the concentration of surfactant increased the repulsive forces. Finally bulk phase experiments were carried out to investigate the efficacy of some commercial sucrose esters in stabilising oil-in-water emulsions.

Sucrose esters have been on the market for a number of years. Their manufacture initially involved the use of dimethylformamide which is a toxic solvent. Hence, they were unsuitable for many applications. Now they are produced by a new reaction technique avoiding the use of toxic solvents and are widely used as emulsifying agents and detergents. The molecule offers the unusual facility that both the degree of esterification and the chain length of the ester group can be altered to obtain a given hydrophile-lipophile balance, so that extensive permutations and combinations are possible. The nontoxic nature of the esters has led to their extensive use in the food industry (1). As sucrose itself stabilizes the conformation of various proteins in aqueous solution, there has been considerable interest in the interactions of sucrose esters (2,3). In our work a number of sucrose esters were synthesised and their properties studied, both in the pure state, and in the presence of aqueous and non-aqueous solvents.

0097–6156/91/0448–0082$06.25/0

The sucrose esters display both thermotropic and lyotropic
liquid crystalline properties.   They resemble the alkyl glycosides
and 1-thioglycosides (4) in forming thermotropic liquid crystalline
phases between room temperature and their melting points and form
lyotropic liquid crystals in water and n-decane.   The phase diagrams
were obtained of sucrose monolaurate, monooleate and dilaurate in
water and of the diesters in n-decane.   Sucrose monolaurate and
monooleate are both fairly water soluble with the formation of
micellar solutions.   The surfactants have sharp cmc's and the
micellar aggregation numbers were determined over the temperature
range 0 to 70°C, using freezing point and vapour pressure methods.
Fundamental studies of their emulsion stabilising properties were
carried out using a light reflectance technique.   The equilibrium
thicknesses of oil-in-water films stabilized by sucrose surfactants
were measured as a function of surfactant and added electrolyte
concentration.   The results were analysed in terms of a simple three
layer oil-water-oil model, which showed that increasing the
concentration of surfactant increased the repulsive forces.

MATERIALS AND METHODS

Materials.   The sucrose esters were prepared using a
transesterification procedure from sucrose and methyl laurate or
oleate in dimethyl formamide as solvent; the monoester was separated
from sucrose and higher sucrose esters by liquid chromatography and
further purified by dialysis (5,6).   Purity was estimated by gas and
thin-layer chromatography as > 99.5%.   The esters as prepared are
mixtures of isomers; the sucrose molecule is most readily esterified
at the primary hydroxyl groups at the 6, 1' and 6' positions (7).
GLC and NMR on methylated derivatives of the monoesters showed that
the isomers 6':6:1' were present in the proportions of 6:3:1.
Sucrose monotallowate was prepared by purifying commercial sucrose
tallowate (Tate and Lyle Industries Ltd.,Reading, U.K.) using silica
gel column chromatography (6).

Experimental.   Phase structures were identified by polarizing
microscopy using a Leitz microscope fitted with a hot-stage.
Hexagonal and lamellar phases were identified by comparing their
textures with literature photomicrographs (8), by their
characteristic conoscopic figures and by low angle X-ray
diffraction.   Cubic phases and micellar solutions may be
distinguished by large differences in viscosity and refractive
discontinuity.   DSC was used to confirm the transitions observed by
optical microscopy.   Homogeneity of the mixtures was achieved using
a vibromixer and repeated centrifugation through a narrow
constriction for more concentrated samples.   Sealed, stirred samples
in a water bath were observed through cross polars, both for heating
and cooling cycles; the phase sequence in more complex regions of
the phase diagram was checked by the microscope penetration
technique.   The aggregation numbers were determined over the
concentration range 0 to 70°C using freezing point and vapour
pressure methods (9).   The osmolality, θ, is related to the freezing
point depression, ΔT, in dilute solution by

$$\theta = \Delta T/K_f, \tag{1}$$

where K_f is the cryoscopic constant, and to the resistance changes, ΔR_i, of the thermistors of the vapour pressure osmometer by

$$\Theta = \Sigma_i \zeta_i \Delta R_i, \tag{2}$$

where the constants ζ_i are obtained by calibration. Solutions were made up by weight and the concentration calculated from the molality using density data. The equilibrium distances between two n-octane droplets in aqueous solutions of the sucrose surfactants were measured as a function of surfactant and of added electrolyte concentration (10). The basis of such a technique is to measure the reflections from the two interfaces of a thin water film sandwiched between two oil droplets using a photomultiplier tube. The intensity of the reflected light is a function of film thickness. The optical arrangement is shown schematically in Figure 1. The beam from a helium-neon laser (λ = 632.8 nm) was reflected by means of an adjustable mirror so that it was incident on the film cell at an angle θ of less than 5°. A beam splitter in the microscope enabled the film thinning process to be monitored visually. The image was focussed at the detector aperture and only the central area of the film was recorded. The intensity of the image was followed continuously on a chart recorder. Actual emulsion stability was assessed by observing creaming and sedimentation behaviour and by studying changes in droplet size with time.

RESULTS

Phase Behaviour. The thermotropic liquid crystalline transitions of the pure sucrose esters are shown in Table I.

Table I. Transition Temperatures of the Sucrose Esters

	Solid - L_β	L_β - L_α	L_α - Isotropic
Sucrose monolaurate	55	138	163
Sucrose monooleate	33	94	154
Sucrose dilaurate	38	81	156
Sucrose dioleate	–	52	89

Sucrose monooleate and dioleate form a gel-like phase at room temperature, but the monolaurate and dilaurate are crystalline solids, which pass into the gel, L_β, phase on heating. On further heating all four esters form the lamellar, L_α, phase.

Thermogravimetric analysis showed that the onset of thermal degradation was around 180°C, but a darkening in colour occurred between 110 and 120°C. X-ray diffraction gave layer spacings of 37 and 43 Å for the monolaurate and monooleate respectively in the lamellar phase.

    Sucrose monolaurate is very soluble in water and the isotropic micellar solution is formed up to concentrations of 57 wt%; above 30

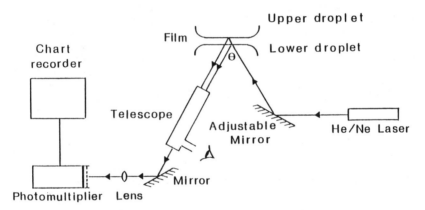

Figure 1. Optical components of the apparatus.

wt% the mixtures show streaming birefringence. The phase diagram is shown in Figure 2. With increasing concentration of surfactant at $20^{\circ}C$ the hexagonal, H_1, phase is formed, followed by the gel phase. Concentrations above 87 wt% formed the lamellar phase on heating. The phase diagram for sucrose monooleate and water is shown in Figure 3. As expected this is not quite so soluble in water as the laurate; the concentrated micellar solutions also showed streaming birefringence. A region of viscous isotropic phase, V_1, is interposed between the hexagonal and gel phases and a much larger lamellar phase region exists.

Sucrose dilaurate is sparingly soluble in both water and n-decane; the phase diagrams are shown in Figures 4 and 5 respectively. Both show regions of lamellar and gel phases. A small area of reversed cubic isotropic phase, V_2, is shown with the hydrocarbon solvent. The longer alkyl chain sucrose dioleate is very soluble in n-decane; this is shown by the phase diagram in Figure 6. Only the lamellar and gel phases are formed.

Sucrose monotallowate is a white powder at ambient temperatures with the fatty acid profile: 35% oleic, 31% palmitic, 25% stearic, small amounts of palmitoleic and linoleic acids. The phase diagram with water is shown in Figure 7. Below $39^{\circ}C$ it is only slightly soluble in water, but above this temperature the solubility increases to 60% and the lamellar phase is formed on adding more water. The low solubility would indicate the presence of considerable amounts of diester, but GLC showed only 5%.

<u>Micellar Aggregation</u>. The cmc of the sucrose monolaurate $(3.39 \times 10^{-4}$ mol dm^{-3} at $25^{\circ}C$) was in agreement with the literature values $(4.0 \times 10^{-4}$ mol dm^{-3} at $25^{\circ}C$ ($\underline{2}$); 3.4×10^{-4} mol dm^{-3} at $27.1^{\circ}C$ $(\underline{12}))$. The cmc of sucrose monooleate was 5.13×10^{-6} mol dm^{-3} at $25^{\circ}C$. Since the surfactants both had a sharp cmc, the thermodynamic data obtained for their solutions was analysed by assuming the micelles to be effectively monodisperse with a single aggregation number independent of concentration. By considering a single-phase micellar solution to be a two component system where the solvent consists of water plus monomer at the cmc and the micelles are the solute, the theory of McMillan and Mayer enables the micelle-micelle interactions to be calculated. The osmotic pressure Π, is given as a power series in the number density, ρ,

$$\Pi/kT = \rho + B_{22}^{*}\rho^2 + B_{222}^{*}\rho^3 + \ldots\ldots \tag{3}$$

Thus, in this model, the non-ideal behaviour of the system is characterized by the virial coefficients B_{22}^{*}, etc. In dilute solution, $B_{22}^{*} \simeq B_{22}^{*o} = -b_{02}^{0}$, the solute-solute cluster integral.

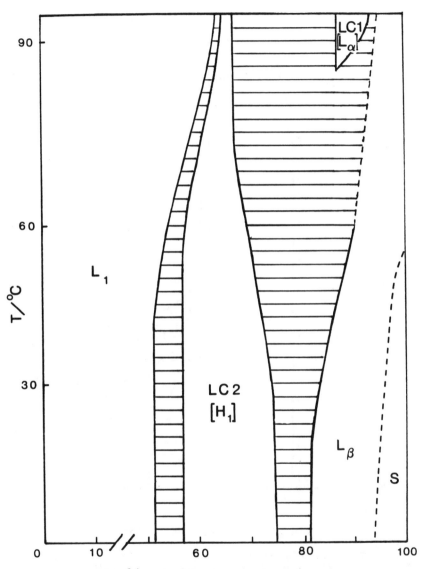

Figure 2.   Phase diagram of the sucrose monolaurate + water
system over the temperature range 0-100°C.   Dotted
lines indicate boundaries not determined accurately.
(Reprinted with permission from ref. 11.
Copyright 1988 American Oil Chemists Society)

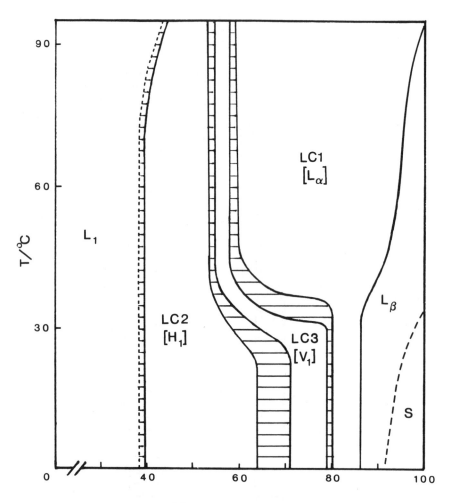

weight % sucrose monooleate

Figure 3.   Phase diagram of the sucrose monooleate + water
system over the temperature range 0-100°C.
Reproduced with permission from reference (11).
(Reprinted with permission from ref. 11.
Copyright 1988 American Oil Chemists Society)

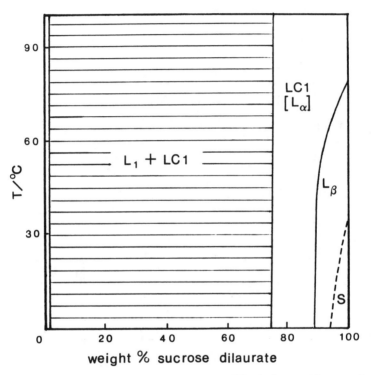

Figure 4. Phase diagram of the sucrose dilaurate + water system over the temperature range 0-100°C. (Reprinted with permission from ref. 11. Copyright 1988 American Oil Chemists Society)

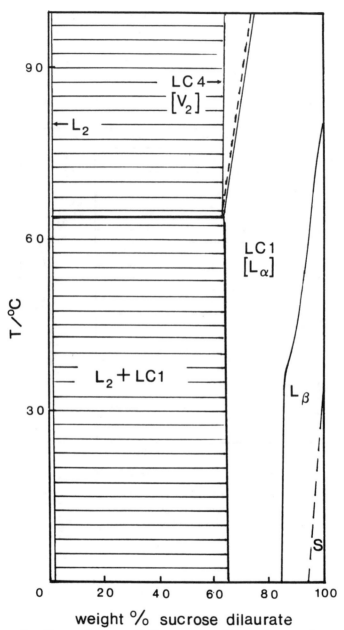

Figure 5.   Phase diagram of the sucrose dilaurate + n-decane
system over the temperature range 0-100°C.
(Reprinted with permission from ref. 11.
Copyright 1988 American Oil Chemists Society)

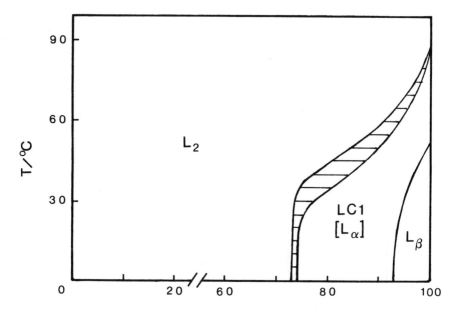

weight % sucrose dioleate

Figure 6.   Phase diagram of the sucrose dioleate + n-decane
             system over the temperature range 0-100°C.
             (Reprinted with permission from ref. 11.
             Copyright 1988 American Oil Chemists Society)

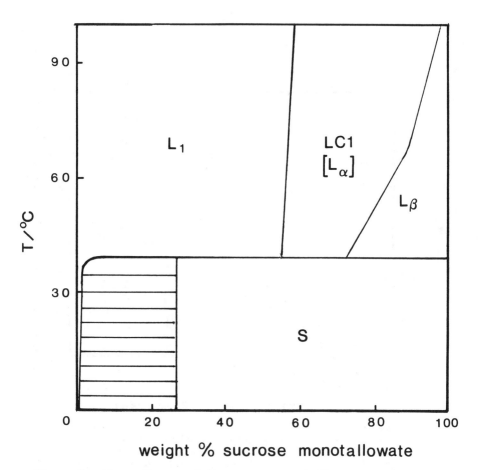

Figure 7.   Phase diagram of the sucrose monotallowate + water
             system over the temperature range 0-100°C.
             (Reprinted with permission from ref. 11.
             Copyright 1988 American Oil Chemists Society)

The micellar molar mass is assumed to be independent of concentration. The osmotic pressure is related to the water activity by

$$\Pi V_1 = -RT \ln a_1 \tag{4}$$

and the osmolality $\Theta = m\phi$ is related to the water activity by

$$\Theta = -\ln a_1/M_1 \tag{5}$$

where M_1 is the molar mass of water.

Thus
$$\Theta(M_1/V_1)/c = 1/M_2 + B_{22}^{*}c/M_2^{2} +\ldots\ldots\ldots \tag{6}$$

where M_2 is the micellar molar mass, which is assumed to be independent of the concentration c. From eqn. (6) a plot of $\Theta(M_1/V_1)/c$ against c has intercept $1/M_2$ and initial slope B_{22}^{*}/M_2^{2}. The molality region studied was $0.01 < m < 0.2$ mol kg^{-1}, well below the concentrations for formation of mesophases. The best fitting polynomials to the data were found to be linear in c at all temperatures. The values of the aggregation number, η, and second virial coefficient, B_{22}^{*}, are given in Table II.

Table II.   Aggregation Numbers, η, and McMillan–Mayer Virial Coefficients, B_{22}^{*}, for Aqueous Solutions of Sucrose Monolaurate and Sucrose Monooleate

T(°C)	Sucrose Monolaurate		Sucrose Monooleate	
	η	B_{22}^{*}/η (cm^3 mol^{-1}x10^3)	η	B_{22}^{*}/η (cm^3 mol^{-1}x10^3)
0	51±4	1.47±0.08	97±8	0.877±0.05
25	52±1	1.40±0.04	99±6	0.764±0.04
40	51±2	1.34±0.05	101±9	0.680±0.05
50	50±2	1.30±0.06	104±10	0.616±0.06
60	54±4	1.12±0.08	96±13	0.572±0.06

Thin Film Studies. A single layer model was used to calculate the
film thickness from the measured intensities, as the paraffin chains
of the surfactant molecules are immersed in the oil phase and will
have almost the same index. For this model

$$\sin^2\phi/2 = \left(\frac{J - J_{min}}{J_{max} - J_{min}}\right)\left(\frac{J_{max} - 1}{J - 1}\right),$$ (7)

where the film thickness, h, is related to ϕ by

$$\phi/2 = 2\Pi n_1 h/\lambda$$ (8)

$J = I_R/I_0$, the ratio of reflected to incident intensity; n_1 is the
film refractive index. The intensity of the incident beam, I_0, was
determined by measuring the reflection from a small plate of quartz
substituted for the film; J_{max} and J_{min} are the last maximum and
minimum intensities recorded as the film thins. The use of this
formula with measurement of J_{max} and J_{min} avoids the need to know
the refractive index of the oil phase; also, as the last term is
effectively unity, measurement of I_0 is unnecessary.

Measurements were carried out on aqueous solutions of sucrose
monolaurate and sucrose monotallowate. The concentration of the
monotallowate in the aqueous phase was kept constant at just above
the cmc (7.41×10^{-3} mol dm^{-3}) and the KCl concentration was varied
between 5×10^{-4} mol dm^{-3} and 7.5×10^{-3} mol dm^{-3}. The results are
shown in Figure 8. It can be seen that, as the concentration of
electrolyte is increased at constant surfactant concentration, the
equilibrium distance decreases, which is consistent with decreasing
repulsive forces. The estimated error in h_e is ± 1 nm. Above an
electrolyte concentration of 7.5×10^{-3} mol dm^{-3} black lunes formed at
the edges of the equilibrium film and it thinned very rapidly to the
Newton Black Film. For the monolaurate two sets of measurements
were made, one at the cmc and the other at one tenth of the cmc. In
each case the effect of electrolyte was the same as for the
monotallowate, but the equilibrium spacings were greater the greater
the concentration of surfactant. Thus increasing the surfactant
concentration increases the repulsive forces and/or reduces the
attractive forces between the n-octane droplets. The intensity
versus time plots as the film thinned showed a continuous variation
with varying electrolyte concentration at constant surfactant
concentration. The plots can be analysed to give information on the
intermolecular forces. The flow of liquid between two surfaces is
given by the Reynolds equation:

$$d(1/h^2)/dt = 4\Delta p/3\eta R^2$$ (9)

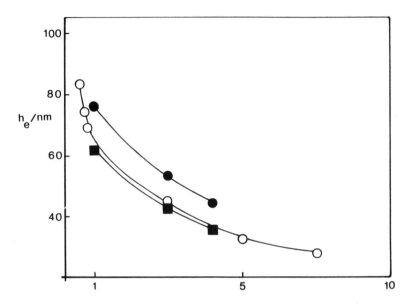

Figure 8.   Equilibrium distance (h_e/nm) as a function of the KCl concentration ($c/10^{-3}$ mol dm^{-3}) for: ○ , sucrose monotallowate, cmc; ■, sucrose monolaurate, cmc/10; ●, sucrose monolaurate, cmc.
(Reprinted with permission from ref. 10. Copyright 1982 Royal Society of Chemistry)

where ΔP is the difference in pressure between the film and the bulk phase, h is the film thickness, η is the viscosity, R is the film radius, ΔP = Π - $Π_\sigma$, where Π is the disjoining pressure and

$$Π_\sigma = 2\sigma r/(r^2 - R^2) \tag{10}$$

The capillary radius, r, was measured using a travelling microscope and the film radius using a grating and graticule.   The Π versus h plots are shown in Figure 9 for sucrose monotallowate.   It can be seen from the curves that the attractive forces increase with increasing electrolyte concentration.   At the highest electrolyte concentration, of 7.5×10^{-3} mol dm^{-3}, some films thinned rapidly to a Newton Black Film.   In general the equilibrium films did not last so long with increasing concentration of electrolyte.   Although the

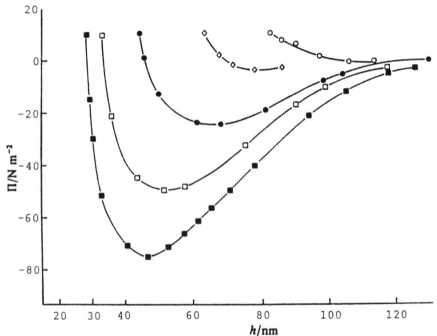

Figure 9. Plots of the disjoining pressure (π/N m^{-2}) against film thickness (h/nm) for sucrose monotallowate at different KCl concentrations (c/10^{-3} mol dm^{-3}): ◯, 0.5; ◇, 1.0; ●, 2.5; □, 5.0; ■, 7.5. (Reprinted with permission from ref. 10. Copyright 1982 Royal Society of Chemistry.)

values for the equilibrium film thickness were larger for sucrose
monolaurate, the films were not so stable as those of the
monotallowate and coalesced to Newton films at a lower concentration
of electrolyte.

Emulsion Stability Studies. The direct emulsion stability studies
were carried out on commercial products, as it was only possible to
synthesize a few grammes of each pure surfactant. These were: TAL
1, sucrose monotallowate - 95% monoester; TAL 2 - 75% monoester; TAL
3 - 50% monoester. The fatty acids were oleic: palmitic: stearic in
the ratio of 35: 31: 25. A range of volume fractions against
n-decane were emulsified. A constant ratio of surfactant in the oil
phase was used throughout. The procedure for emulsification was
standardized. The emulsions were left in screw-top vials in a 25°C
air thermostat. Photographs were taken to show the stability order
with respect to sedimentation and creaming over a period of several
weeks. As expected the ester with the highest percentage of
monoester formed stable emulsions up to a volume fraction of 0.6
over a week; higher volume fractions creamed rapidly or
spontaneously inverted. One of the problems with these esters was
their low solubilities; above 1% the solutions are turbid and the
2-phase region of the phase diagram is entered. Calculations based
on droplet size showed that 1% by weight was unlikely to produce a
close-packed monolayer at the interface.

DISCUSSION

The thermotropic behaviour of the sucrose esters is analogous to
that of octyl, nonyl and decyl glucopyranosides (4); crystals of
these have transitions to an intermediate followed by a "smectic"
phase on heating. Also in the lamellar and gel phases there is a
similar tendency to form homeotropic textures (11). By analogy with
ionic surfactants, the dialkyl esters would be predicted to form L_{α},
V_2 or H_2 mesophases. Aqueous solutions of the dialkyl
polyoxyethylene surfactant, $(C_{10}H_{21})_2CHCH_2(OCH_2CH_2)_{10}OH$, show an
extensive L_{α} phase changing to V_2 above 91°C. Sucrose dilaurate and
dioleate show lower transition temperatures both for L_{β} to L_{α} and L_{α}
to isotropic liquid transitions than the monoalkyl esters. The
phase diagram of sucrose dilaurate with water shows a more extensive
L_{α} phase than that of the monolaurate, but no sign of the hexagonal
phase; also the V_2 phase is only shown with n-decane.

The factors determining the effect of headgroup size and alkyl
chain length on the type of liquid crystal formed are the same as
those dictating micelle shape (13). Possible micelle shapes are
spheres, rods or discs. The volume per alkyl chain, v, surface area
per chain, A_c, and the longest extension of the alkyl chain, l_t, are
related by:

$$A_c > 3v/l_t \quad \text{(sphere)}$$
$$A_c > 2vl_t \quad \text{(rod)}$$
$$A_c > vl_t \quad \text{(disc)}$$

where $l_t/\text{Å} = 1.5 + 1.265\ n_c$ (n_c is the number of carbon atoms embedded in the micelle core; for sucrose monolaurate $n_c = 11$) (14); v is calculated from the density of undecane (15). The limiting values for sucrose monolaurate are

$$A_c > 68.6\ \text{Å}^2 \quad \text{(spheres)}$$
$$68.6 > A_c > 45.7\ \text{Å}^2 \quad \text{(rods)}$$
$$45.7 > A_c > 22.9\ \text{Å}^2 \quad \text{(bilayers)}$$
$$22.9 > A_c \quad \text{(reversed phases)}$$

Thus the transitions would be expected to occur in the order: cubic (I_1) -> hexagonal (H_1) -> cubic (V_1) -> lamellar (L_α) -> reversed phases. Considering the micelles as hard-core particles, the phase transitions will be determined by packing constraints. The limiting volume fractions are 0.74 for face-centered cubic (spherical micelles) and 0.91 for the hexagonal phase (rod-shaped micelles). The limiting volume fractions are found experimentally to be within 70 to 80% of the close-packed values. The phase diagrams are consistent with this general picture. The streaming birefringence shown by surfactant-rich isotropic solutions of the monolaurate and monooleate are considered to indicate the presence of rod-shaped micelles and hence the incipient formation of the hexagonal phase. Sucrose monolaurate shows a more extensive hexagonal phase and less lamellar phase than the monooleate of longer alkyl chain length. This is similar to the behaviour of the polyoxyethylene surfactants: comparison of $C_{12}EO_6$ with $C_{10}EO_6$ and of $C_{16}EO_8$ with $C_{12}EO_8$ shows that the former exhibit more lamellar and less hexagonal phase than the latter. However for the monoglycerides increasing the alkyl chain length decreases the stability of the lamellar phase (16). Most of the polyoxyethylene surfactants show a lower consolute temperature, implying that the headgroup hydration decreases with increasing temperature for a given alkyl chain length; this would favour the lamellar phase at higher temperatures and is supported by the phase diagrams. The sucrose surfactants showed no signs of a cloud point in the temperature range studied, consistent with the strongly hydrophilic character of the sucrose head group.

     For surfactants with a fairly high and ill-defined value of the cmc, such as octylmethylsulphoxide, it is possible to study the changes of the thermodynamic properties on micellization by using a mass action model to analyse the data (17). However, when the cmc is low, it is impossible to obtain accurate thermodynamic data in the premicellar region. Recently, the micellar properties of the n-alkyl polyethyleneoxide surfactants, C_nEO_j, which have low values of the cmc, but also show a lower consolute temperature or cloud point, have been studied by neutron scattering, NMR and dynamic light scattering (18-20). However, the fundamental issue of the effect of temperature in causing predominantly micellar growth or increasing the aggregation of small micelles has not been resolved. Sucrose monolaurate and monooleate also have very low values of the cmc, but differ from the C_nEO_j series in two respects; the formation of liquid crystalline phases in the pure state and the absence of a cloud point below 100°.

The aggregation number of the monolaurate is considerably less than that of the monooleate at all temperatures studied. This can be attributed to the larger hydrocarbon length of the oleyl group. Similar behaviour is shown by the n-alkylpolyethylene oxide surfactants; for example at $25^\circ C$ the aggregation number is 400 for $C_{12}EO_6$ but 2430 for $C_{16}EO_6$ (9). Also the η values for the sucrose monesters are less than those of the C_nEO_j series with comparable chain length and head group size; for sucrose monolaurate η = 50, whereas for $C_{12}EO_6$ η = 400 at $25^\circ C$. This is consistent with an aggregation number decreasing with increasing hydrophilic nature of the head group. Thus smaller micelles are favoured for the sucrose esters. The strongly hydrophilic nature of the sucrose head group also explains the lack of a cloud point below $100^\circ C$.

The previous equation (6) may be written as

$$\Pi/RTc = 1/M_2 + B_{22}^{*}cM_2^{2}+ \ \ldots\ldots\ (11)$$

The Π/c versus c plots of both sucrose monoesters show a consistent trend. Increasing temperature steadily decreases the initial positive slope of the curves, which reflects steadily decreasing values of B_{22}^{*}, the virial coefficient for micelle-micelle interaction. A negative value would indicate micelle-micelle attraction with possible aggregation. This indicates that a cloud point may exist above $100^\circ C$. For $C_{12}EO_6$, B_{22} is positive at $18^\circ C$, zero at $20^\circ C$ and negative at $45^\circ C$ as expected for a cloud point at $50^\circ C$. This is an important difference between the sucrose and polyethylene oxide head groups. The ethylene oxide chains form a flexible coiled head group, a helical coil, which can alter size and shape, whereas the sucrose head group presents a relatively rigid structure.

Thin liquid films are often used as models for studying emulsion stability as their equilibrium thickness is determined by the same forces. The stability of emulsions depends on two processes. Firstly, the reversible flocculation of the droplets of the dispersed oil droplets, with formation of a thick "Common Black Film". The thickness of this film is determined by the opposition of van der Waals attractive and electrostatic repulsive forces. These films are formed at low ionic strength, the long range repulsive forces hinder aggregation. Secondly, in solutions of higher ionic strength, a second thinning process may occur and the droplets "coalesce" to form the very thin "Newton Black Films" Here the van der Waals attractive forces are balanced by the repulsive steric forces. Our measurements were restricted to the thicker "Common Black Films".

The energy of interaction between the two oil droplets has 3 components:

1) the overlap of the electrical double layer gives a repulsive pressure, Π_R;

2) the van der Waals attractive pressure, Π_A;

3) steric stabilisation produced by molecular adsorption at the surface, Π_S. Their sum is the disjoining pressure, Π_D. Thus

$$\Pi_D = \Pi_R + \Pi_A + \Pi_S \qquad (12)$$

The contribution of the various components as a function of film thickness is shown in Figure 10. At the equilibrium thickness the capillary pressure, Π_σ, which tends to thin the film is equal to the disjoining pressure:

$$\Pi_\sigma = \Pi_D \qquad (13)$$

Assuming that the double layer potential, ψ_o, was independent of the electrolyte concentration, a theoretical analysis showed that ψ_o, and hence the repulsive forces, increase with increasing concentration of surfactant. However this result is different from that observed with the nonionic surfactant NP20 (nonylphenylpolyethyleneoxide-20). In both this work and that of Sonntag et al (21) it was found that increasing the concentration of this surfactant from one tenth of the cmc to the cmc reduced the equilibrium spacings of the n-octane droplets.

It was hoped to correlate the distance apart of the oil droplets with emulsion stability. It must be remembered that it is necessary to use a theoretical model to interpret the results as even if h_e(cmc) > h_e(cmc/10), then, since Π_σ(cmc) < Π_σ(cmc/10), it does not necessarily mean that the repulsive forces are greater. It seems that the repulsive forces may be increased or reduced by increasing the surfactant concentration depending on the system studied. The values of h_e also only reflect the equilibrium situation. In a dynamic situation the rate at which the surfactant redistributes itself at the surface as the film thins must play a part. In this connexion, it was observed that the sucrose esters thinned to a Newton Black film in a characteristic manner that was quite different from the thinning of NP20. For the sucrose esters a black lune formed at the edge and then rapidly spread to cover the whole film, whereas for NP20 black spots appeared in the middle which then enlarged to embrace the whole film. The difference probably reflects different surface viscoelasticities of the films.

CONCLUSIONS

Sucrose monoesters of long chain acids are very soluble in water and the increasing interdroplet distance with increasing surfactant

DISJOINING PRESSURE −FILM THICKNESS

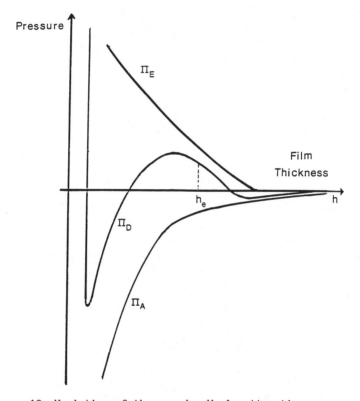

Figure 10. Variation of the van der Waals attractive pressure, the electrostatic repulsive pressure and the disjoining pressure with film thickness.

concentration implies that they should be excellent emulsifiers. Experiments will be carried out to determine the surface rheological parameters of the interfacial film as this undoubtedly affects long term emulsion stability. The problem is the insolubility of commercial sucrose esters, because of the presence of diesters and higher esters; simple procedures to remove the higher esters are worth considering.

LITERATURE CITED

1. Yamada, T; Kawase, N; Ogimoto, K. Yukagaku, 1980, 29, 543.
2. Makino, S; Ogimoto, S; Koga, S. Agric. Biol. Chem. 1983, 47, 319.
3. Tomida, M; Kondo, Y; Moriyama, R; Machida, H; Makino, S. Biochim. Biophys. Acta, 1988, 943, 493.
4. Jeffrey, G. A.; Battacharjee, S. J. Am. Oil Chem. Soc. 1983, 60, 1908.
5. Osipow, L. I.; Snell, F. D.; York, W. C. ; Finchler, A. Ind. Eng. Chem. 1956, 48, 1459.
6. Gupta, R. K.; James, K.; Smith, F. J. J. Am. Oil Chem. Soc. 1983, 60, 1908.
7. York, W. C.; Finchler, A.; Osipow, L.; Snell, F. D. J. Am. Oil Chem. Soc. 1956, 33, 424.
8. Rosevear, F. B. J. Am. Oil Chem. Soc. 1954, 31, 628.
9. Herrington, T. M.; Sahi, S. S. Colloids and Surfaces, 1986, 17, 103.
10. Herrington, T. M.; Midmore, B. M.; Sahi, S. S. J. Chem. Soc., Faraday Trans. 1, 1982, 78, 2711.
11. Herrington, T. M.; Sahi, S. S. J. Am. Oil Chem. Soc. 1988, 65, 1677.
12. Osipow, L.; Snell, F. D.; Hickson, J. J. Am. Oil Chem. Soc. 1959, 35, 127.
13. Tiddy, G. J. T.; Walsh, M. F. Studies in Theoretical Chemistry, 1983, 79, 975.
14. Tanford, C. The Hydrophobic Effect; Wiley-Interscience: New York, 1980; p 52.
15. Hall, K. R. Selected Values of Properties of Hydrocarbons and Related Compounds; Texas A & M University: Texas, 1983; A-85.
16. Lutton, E. S. J. Am. Oil Chem. Soc. 1965, 42, 1068.
17. Corkill, J. M.; Walker, T. J. Colloid Interface Science 1972, 39, 62.
18. Nilsonn, P. G.; Wennerstrom, H.; Lindman, B. J. Phys. Chem. 1983, 87, 1377.
19. Brown, W.; Johnsen, R.; Stilbs, P.; Lindman, B. J. Phys. Chem. 1983, 87, 4548.
20. Zulauf, M.; Rosenbusch, J. P. J. Phys. Chem. 1983, 87, 856.
21. Sonntag, H.; Netzel, J.; Unterberger, B. Spec. Discuss. Faraday Soc. 1970, 1, 57.

RECEIVED September 25, 1990

Chapter 8

Liposarcosine-Based Polymerizable and Polymeric Surfactants

Bernard Gallot

Laboratoire des Materiaux Organiques, Centre National de la Recherche
Scientifique, B.P. 24, 69390 Vernaison, France

Polymerizable liposarcosine surfactants have been
synthesized from aliphatic α,ω-diamines by fixing a
polymerizable group at one end and a sarcosine
chain with various degrees of polymerization at the
other. Then polymerizable liposarcosine have been
transformed into comb copolymers by radical
homopolymerization or radical copolymerization with
long aliphatic chain alkyl acrylamides to obtain
polymeric surfactants with a large range of
hydrophilic lipophilic balance. Polymerizable and
polymeric surfactants exhibit both lyotropic and
emulsifying properties which have been analyzed.

Amphiphilic lipopeptides $H\text{-}(CH_2)_n\text{-}NH\text{-}(AA)_p$ formed by a
hydrophobic paraffinic chain containing from 12 to 18 carbon
atoms and a hydrophilic peptidic chain in which the repeating
unit (AA) is one of the following amino acids: sarcosine, serine,
lysine, glutamic acid, hydroxyethylglutamine and hydroxy-
propylglutamine exhibit both lyotropic (1-3) and emulsifying
properties (4-6). In order to fix the structure of such mesophases
and emulsions it is interesting to prepare them with
polymerizable surfactants that can be postpolymerized. A good
method to obtain at first polymerizable surfactants and then
polymeric surfactants consists in replacing the terminal methyl
group of the paraffinic chain of the lipopeptides by a
polymerizable group.

Before undertaking the synthesis of polymerizable
surfactants three parameters have to be determined: the nature

of the amino acid, the length of the hydrophobic chain and the nature of the polymerizable group. Our choice is based on the ability of different polymerizable groups to be easily polymerized and on the knowledge<of the properties of amphiphilic lipopeptides. Among lipopeptides liposarcosines with a hydrophobic chain containing from 12 to 18 carbon atoms exhibit the best emulsifying properties (4,5). Furthermore acrylate, methacrylate, acrylamide and methacryl-amide groups easily polymerize by radical polymerization, so the best polymerizable surfactants would be liposarcosines with the following general formula:

$$H_2C= CR-CO-Y-(CH_2)_{12}-NH-[CO-CH_2-N(CH_3)]_p-H$$
with R= H or CH_3 and Y= NH or O

In this paper we describe the synthesis and properties of such liposarcosine based polymerizable and polymeric surfactants.

EXPERIMENTAL METHODS
Synthesis of the Polymerizable Amine.
Protection of the Amine Function. Ditert-butyldicarbonate (0.1 mole) is reacted at room temperature with 0.1 mole of 1,12-dodecyldiamine in solution in 300 ml. of a water/methanol mixture (50/50, v/v) and in presence of a sufficient quantity of NaOH to maintain a pH higher than 8. After 12 hours of reaction the $Boc-NH-(CH_2)_{12}-NH_2$ is precipitated by addition of water and purified. The protected amine is separated at first from the starting amine by extraction with THF, then from diacyl derivative by chromatography on a silica column with methanol containing 1% of ammoniac as eluent.
Fixation of the Polymerizable Group. Acryloyl chloride (0.11 mole) in solution in 200 ml. of THF containing 0.11 mole of triethylamine is added dropwise to a solution of 0.1 mole of $Boc-NH-(CH_2)_{12}-NH_2$ in 300 ml. THF. The precipitate of triethylaminehydrochloride is separated by filtration and the desired product is precipitated by addition of water.
Elimination of the Protective Group. The protective group (Boc) is eliminated by the action of an excess of hydrobromic acid (HBr,4N) on the acyl derivative in acetic acid solution. After several hours, hydrobromic acid and acetic acid are eliminated by distillation followed by lyophilization. The salt of the polymerizable amine is dissolved in water and neutralized by addition of NaOH.

At each step of the synthesis the product obtained is characterized by Infrared Spectroscopy and its purity is verified by thin layer chromatography on silica gel with methanol containing 1% ammonia as eluent.

Synthesis of Polymerizable Surfactants
<u>N-carboxyanhydride Method.</u> Sarcosine N-carboxyanhydride (0.03 mole), prepared by the action of phosgene on sarcosine (7), is reacted with 0.01 mole of the polymerizable amine in solution in 250 ml. of THF, at room temperature, under agitation for one hour The solvent is evaporated and the residue washed by ice-cold acetone and lyophilized.

The average number of amino acid residues coupled to the polymerizable amine is determined by titration of the terminal amine group by perchloric acid in acetic acid solution. A value of 2.8 is found.

<u>Coupling Method.</u> The polymerizable amine (0.01 mole), 0.01 mole of t-butyloxycarbonylsarcosine (SarBoc) (8-9), 0.01 mole of N-hydroxysuccinimide and then 0.01 mole of dicyclohexyl-carbodiimide are dissolved in 250 ml. of THF. After 24 hours of agitation at room temperature, the dicyclohexylurea precipitate is eliminated by filtration and the N-protected polymerizable liposarcosine is precipitated by addition of one volume of water. The amine group of the sarcosine is then unblocked by successive action of HCl (5N) in ethyl ether solution and NaOH (1N) in methanol solution. The polymerizable liposarcosine is purified by chromatography on a silica column and eluting with methanol containing 1% ammoniac.

After each step of the synthesis, the product obtained is characterized by Infrared Spectroscopy and its purity is verified by thin layer chromatography on silica gel with 1% ammonia in methanol eluent.

Synthesis of Polymeric Surfactants
<u>Homopolymerization.</u> Polymerizable liposarcosine (0.01 mole) with p=2.8 is dissolved in 150 ml. of chloroform and 70 mg. of AIBN is added. After elimination of oxygen by bubbling nitrogen into the solution, the system is maintained at 60°C for 48 hours. Then the comb polymer is precipitated by adding a large volume of water and dried.

<u>Copolymerization.</u> Polymerizable liposarcosine (0.01 mol) with p=2.8 and 0.01 mole of myristoylacrylamide are dissolved in 150 ml. of chloroform and 70 mg. of AIBN is added; after elimination

of oxygen by bubbling nitrogen into the solution, the system is maintained at 60°C for 48 hours. Then the comb copolymer is precipitated by adding a large volume of water and dried. The composition of the random copolymer is determined by titration of the terminal amine group of sarcosine by perchloric acid in acetic acid solution.

X-ray Diffraction Studies. Polymeric surfactants are dissolved in a small excess of water. When total homogeneity is achieved, the desired concentration is obtained by slow evaporation at room temperature. The sample is then left at room temperature in tight cells to reach equilibrium. At last dry polymeric surfactants and their water concentrated solutions are studied by X-ray diffraction as a function of temperature using a Guinier type focussing camera equipped with a bent quartz monochromator giving a linear collimation and operating under vacuum (10).

Preparation of Emulsions   The mixture oil-surfactant is heated to 70°C under agitation with a Ystral Gmbh type X1020 apparatus for complete homogeneity.   A   small amount of polymerization inhibitor is added to the system in the case of polymerizable surfactants. The system is then cooled to 50°C and water is added while agitation is maintained until total homogeneity is achieved (11).

Stability of Emulsions. The stability of emulsions is determined by following, as a function of time, the variation of the emulsified volume at fixed temperatures. Emulsions are considered stable if after 60 days the emulsified volume is still 100%.

RESULTS

Synthesis of Polymerizable Surfactants
Amphiphilic lipopeptides were synthesized from fatty amines. The primary amine function allows linking an amino acid to an aliphatic chain through an amide bond, or to initiate the polymerization of the N-carboxyanhydride (NCA) of the amino acid (1,4,5). To synthesize polymerizable lipopeptides, the starting lipids have to be α,ω-bifunctional lipids with an amine function to link the peptidic chain and a second function to link the polymerizable group. The nature of the second functional group (NH_2 or OH) is determined by the type of polymerizable

group wanted (acrylamide or acrylic) But, as the longest aliphatic α,ω-diamine and α,ω-aminoalcohol commercially available contains respectively 12 and 5 carbon atoms, we used at first, as bifunctional lipid, the 1,12-dodecylamine to have a hydrophobic chain similar to that of the lipopeptides exhibiting the best emulsifying and foaming properties (4-6).

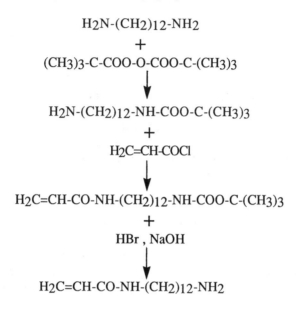

$$H_2N\text{-}(CH_2)_{12}\text{-}NH_2$$
$$+$$
$$(CH_3)_3\text{-}C\text{-}COO\text{-}O\text{-}COO\text{-}C\text{-}(CH_3)_3$$
$$\downarrow$$
$$H_2N\text{-}(CH_2)_{12}\text{-}NH\text{-}COO\text{-}C\text{-}(CH_3)_3$$
$$+$$
$$H_2C\text{=}CH\text{-}COCl$$
$$\downarrow$$
$$H_2C\text{=}CH\text{-}CO\text{-}NH\text{-}(CH_2)_{12}\text{-}NH\text{-}COO\text{-}C\text{-}(CH_3)_3$$
$$+$$
$$HBr , NaOH$$
$$\downarrow$$
$$H_2C\text{=}CH\text{-}CO\text{-}NH\text{-}(CH_2)_{12}\text{-}NH_2$$

Method   a:

$$H_2C\text{=}CH\text{-}CO\text{-}NH\text{-}(CH_2)_{12}\text{-}NH_2$$
$$+$$
$$SarBoc$$
$$\downarrow$$
$$H_2C\text{=}CH\text{-}CO\text{-}NH\text{-}(CH_2)_{12}\text{-}NH\text{-}SarBoc$$
$$+$$
$$HBr , NaOH$$
$$\downarrow$$
$$H_2C\text{=}CH\text{-}CO\text{-}NH\text{-}(CH_2)_{12}\text{-}NH\text{-}CO\text{-}CH_2NH(CH_3)$$

Method b:

$$H_2C=CH-CO-NH-(CH_2)_{12}-NH_2$$

$$+$$

p. SarNCA

$$\downarrow$$

$$H_2C=CH-CO-NH-(CH_2)_{12}-NH-[CO-CH_2-N(CH_3)]_p-H$$

Starting from the 1,12-dodecyldiamine, the polymerizable surfactant is synthesized in four steps: protection of one amine function by a tertiobutyloxycarbonyl group (Boc) through the action of ditertiobutylpyrocarbonate (diBoc), fixation of the polymerizable group through the action of acryloyl- or methacryloyl-chloride, elimination of the protecting group through the action of hydrogen bromide and sodium hydroxide, fixation of the peptidic chain through a coupling reaction between the polymerizable amine and the N-protected sarcosine (SarBoc) followed by the elimination of the protecting group (see method a on the synthesis scheme), or through the polymerization of the N-carboxyanhydride of sarcosine (SarNCA) at the end of the polymerizable amine (see method b on the synthesis scheme). The coupling method was used to obtain a polymerizable liposarcosine with one or two residues of sarcosine, and the NCA method to obtain polymerizable liposarcosines with more than two residues of sarcosine.

Synthesis of Polymeric Surfactants.
In order to obtain a large domain of HLB for polymeric surfactants, polymerizable liposarcosines have been homopolymerized and copolymerized with long aliphatic chain alkylacrylamide. The polymerization conditions have been determined by the respective solubilities of polymerizable liposarcosines and alkylacrylamides, as the synthesis of all polymers has to be performed in the same conditions to allow a direct comparison of the properties of homo and copolymers. After trying different polymerization conditions (polar and non polar solvents, persulfates, peroxydes and azoinitiators) we found that the best polymerization conditions (conditions that allow the synthesis of homopolymers and copolymers of any composition with a good yield) are the following: chloroform solution,

temperature of 60°C, azobisisobutyronitrile (AIBN) initiator, absence of oxygen.

Liquid Crystalline Properties

X-ray diffraction studies showed that liposarcosine based comb copolymers exhibit both thermotropic and lyotropic properties (12). The mesophases observed are Smectic A and Nematic, and the domain of stability of the two types of mesophases is governed by the temperature and the water concentration of the systems (12). In order to illustrate the lyotropic behaviour of the liposarcosine based comb copolymers we have plotted in Figure 1 the variation of the smectic layer thickness **d** as a function of the water content of the system for a comb copolymer with an average degree of polymerization of the sarcosine p=2.8. One can see that the smectic layer thickness **d** increases when the water content of the mesophase increases.

Emulsifying Properties   We describe successively the

emulsifying properties of polymerizable liposarcosine   and liposarcosine based comb copolymers.

Polymerizable Surfactants. The emulsifying properties of

alkylacrylamide liposarcosines are similar to that of aliphatic liposarcosine already described (4-6). They easily give stable emulsions with a lot of oils such as isopropylmyristate, butylstearate, decane, dodecane, mygliol, cosbiol etc. The emulsions vary from a fluid milk to a thick cream, depending upon the nature of the oil, the ratio oil/water, the degree of polymerization p of the sarcosine in the surfactant and the concentration of surfactant.

All the emulsions are of the oil in water (O/W) type   as shown by the dilution method (a droplet of emulsion immediately disappears in water but falls at the bottom of a vessel containing water). The selective dye method (a droplet of emulsion deposited on a pulverised mixture of an oil soluble red dye and a water soluble green dye   gives a red spot surrounded by a green halo) and the conductivity method (a conductivity of 10^{-3} Ω^{-1} cm^{-1} was found for emulsions prepared with dilute KCl solutions).

For a given oil, the domain of stability of the emulsions decreases when the degree of polymerization of the sarcosine increases, as already observed for aliphatic liposarcosines (6).On the contrary the domain of stability of the O/W emulsions decreases when the polarity of the oil decreases.

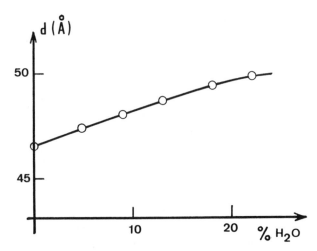

Figure 1.   Example of variation of the smectic layer thickness
d as a function of the water content of the system.

Figure 2 illustrates quantitatively the behaviour of
acrylamide liposarcosines in the case of a liposarcosine of
formula: $CH_2= CH-CO-NH-(CH_2)12-NH-CO-CH_2-NH(CH_3)$
The stability domain of the O/W emulsions has been established
by studying systems with increasing amounts of oil and
decreasing amounts of surfactant. We determined at first the
maximum amount of oil that can be incorporated in stable O/W
emulsions.We determined then, for each oil content of O/W
emulsions, the minimum amount of surfactant giving stable
emulsions.
Polymeric Surfactants. In the case of polymerizable surfactants
the HLB can be modified only by changing the hydrophilicity of
the surfactant by varying the degree of polymerization of the
sarcosine. In the case of polymeric surfactants one can modify at
will both the hydrophilic and the hydrophobic character of the
surfactant. One can increase the hydrophilicity of the polymer by
increasing the degree of polymerization p of the sarcosine, and
the hydrophobicity of the polymer by copolymerization of any
acrylamide liposarcosine with increasing amounts of long
aliphatic alkyl acrylamide.
     To obtain polymeric surfactants exhibiting a large range of
HLB, we used acrylamide-liposarcosines with different degrees
of polymerization p of the sarcosine. We homopolymerized them
and copolymerized them with different amounts of aliphatic
alkylacrylamide. Table I gives the characteristics of two series
of homopolymers and copolymers studied.

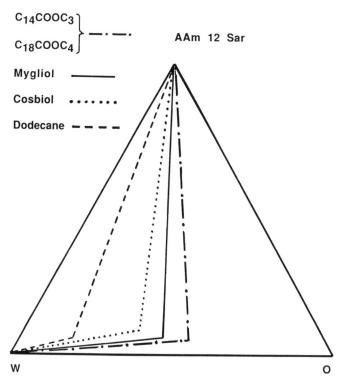

C₁₄COOC₃
C₁₈COOC₄   — · — ·   AAm 12 Sar

Mygliol   ———

Cosbiol   · · · · · ·

Dodecane   — — —

W   O

Figure 2.   Influence of the nature of the oil on the domain of stability of O/W emulsions obtained with the polymerizable surfactant AAm12Sar.

TABLE I.   Molecular Characteristics of Some Comb Polymers Studied

POLYMER	P	b/a
AAmSar6	6.5	0
AAmSar6-C14A	6.5	1.6
AAmSar6-C14B	6.5	2.5
AAmSar3	2.8	0
AAmSar3-C14A	2.8	1
AAmSar3-C14B	2.8	2

b/a: molar ratio of a and b in the polymer
a: $CH_2=CH-CO-NH-(CH_2)_{12}-NH-[CO-CH_2-N(CH_3)]_p-H$
b: $CH_2=CH-CO-NH-(CH_2)_{14}-H$

The emulsifying properties of the polymeric surfactants have been studied with two types of oil/water systems, namely isopropylmyristate/water and dodecane/water.

Homopolymers AAmSar6 and AAmSar3 emulsify with difficulty isopropylmyristate/water systems and are not able to emulsify (at any surfactant concentration) dodecane/water systems.

Copolymers AAmSar6-C14A and AAmSar3-C14A emulsify isopropylmyristate/water systems with smaller amounts of surfactant than the corresponding homopolymers, but are not able to emulsify dodecane/water systems.

Copolymers AAmSar6-C14B and AAmSar3-C14B are richer in myristoylacrylamide and able to emulsify both systems isopropylmyristate/water and dodecane/water. The amounts of polymeric surfactants necessary to obtain stable emulsions are three times smaller for isopropylmyristate than for dodecane. Furthermore the copolymer content of myristoylacrylamide necessary to obtain stable emulsions seems to increase with increasing the sarcosine degree of polymerization.

Conclusions.
Liposarcosine based comb copolymers exhibit interesting emulsifying properties. Much work has to be performed to optimize their properties and to relate them with the molecular characteristics of the polymers, namely: the nature of the main chain, the number n of carbon atoms of the paraffinic chain and the degree of polymerization p of the sarcosine in the liposarcosine side chains, the length of the alkyl chain of the acrylic, methacrylic, acrylamid or methacrylamid comonomer and the ratio liposarcosine/comonomer.

Furthermore, as other amino-acids such as lysine, glutamic acid, aspartic acid etc;, also give raise to comb copolymers exhibiting emulsifying properties (13), comparative studies of these copolymers with liposarcosine based copolymers will be performed in the near future.

LITERATURE CITED

1. Douy, A.; Gallot, B. Makromol.Chem. 1986, 187, 465
2. Gallot, B.; Douy, A.; Haj Hassan, H. Mol.Cryst.Liq.Cryst. 1987, 153, 465
3. Gallot, B.; Haj Hassan, H. Mol.Cryst.Liq.Cryst. 1989, 170, 89

4. Gallot, B.; Douy, A. French Patent 82 159 76, 1982; Chem.Abstr. 1984, 171762h

5. Gallot, B.; Douy, A. U.S.Patent 4 600 526 , 1986

6. Gallot, B.; Haj Hassan, H. In *Polymer Association Structures: Liquid Crystals and Microemulsions* ; El-Nokaly, M.A., Ed.; ACS Symposium Series 384, New Orleans, 1989, pp 116-128.

7. Fuller,W.D.; Goodman, E.; Werlander, M.S. Biopolymers 1976, 15,1869

8. Moroder, L.; Hallett, S.; Wunch, E.; Keller; O.;Wersin, G. Z.Physiol.Chem., 1976, 357, 1651

9. Keller, O. Org.Synth. 1985, 63, 160

10. Douy, A.; Mayer, R.; Rossi, J.; Gallot, B. Mol.Cryst.Liq.Cryst. 1969, 7, 103

11. Haj Hassan, H. Ph.D. Thesis, University of Orléans at Orléans, France, June 1987.

12. Gallot, B.; Douy, A. Mol.Cryst.Liq.Cryst. 1987, 153, 367

13. Gallot, B.; Douy, A. U.S. Patent 4 859 753 , 1989

RECEIVED August 28, 1990

Chapter 9

Competitive Adsorption and Protein–Surfactant Interactions in Oil-in-Water Emulsions

Eric Dickinson

Procter Department of Food Science, University of Leeds, Leeds LS2 9JT, United Kingdom

Competitive adsorption in oil–in–water emulsions has been investigated for binary protein/protein and protein/surfactant systems. Analysis of aqueous phase composition in emulsions (10 wt % n-tetradecane, pH 7) indicates fast reversible adsorption for α_{s1}–casein/β–casein, but slow irreversible adsorption for β–casein/α–lactalbumin, β–casein/β–lactoglobulin and α–lactalbumin/β–lactoglobulin. Surface viscosity data at the planar oil–water interface give further information on milk protein competitive adsorption. Interfacial tensions and viscosities are reported for a mixture of a non–ionic surfactant, octaoxyethylene dodecyl ether ($C_{12}E_8$), and sodium caseinate; and stability data are presented for emulsions made with $C_{12}E_8$ + caseinate. Differences between $C_{12}E_8$ and the anionic surfactant, sodium dodecyl sulphate (SDS), are noted.

Most food emulsions are stabilized by proteins, either alone or in combination with low–molecular–weight surfactants (lipids or their derivatives) or high–molecular–weight polymers (polysaccharide hydrocolloids) (1–4). The distribution of these different kinds of molecules between the droplet surface and the bulk phases is an important factor controlling the formation, stability and texture of edible oil–in–water emulsions such as cream liqueurs, salad dressings, or ice–cream. What affects the distribution is (a) competitive adsorption between different proteins and between the proteins and the surfactants and (b) interactions between proteins and the other components at the oil–water interface. It is clear that interfacial protein–polysaccharide interactions can have a significant effect on emulsion stability (3,5), but for reasons of brevity this paper is concerned only with interfacial protein–protein and protein–surfactant interactions.

The proteins chosen for study are the milk proteins— α_{s1}–casein, β–casein, α–lactalbumin and β–lactoglobulin. We report results of experiments designed to quantify the extent to which one pure milk protein A is able to displace another pure milk protein B from the

0097–6156/91/0448–0114$06.00/0

oil—water interface. An adsorbed film of protein B is established at
the emulsion droplet surface, or at a macroscopic planar oil—water
interface, and then protein A is introduced into the bulk aqueous
phase. Time-dependent competitive adsorption is monitored by
following the change in protein composition of the serum phase (in the
case of the emulsion) or the change in surface viscosity of the
adsorbed film (in the case of the planar interface). Experiments are
carried out at neutral pH with n-tetradecane as the oil phase. Data
are also presented for a binary system composed of the proteinaceous
food emulsifier, sodium caseinate, and the non-ionic water-soluble
surfactant, octaoxyethylene dodecyl ether ($C_{12}E_8$).

Experimental

Materials. Samples of α_{s1}-casein and β-casein were prepared from
whole milk using standard procedures (6,7). The α-lactalbumin and
AnalaR-grade n-tetradecane were obtained from Sigma Chemicals (St.
Louis, MO). The β-lactoglobulin and sodium dodecyl sulphate were
obtained from BDH Chemicals (Poole, UK). High purity $C_{12}E_8$ was
obtained from Nikko Chemicals (Tokyo, Japan). Sodium caseinate was
obtained from the Scottish Milk Marketing Board (Renfrewshire, UK).
Buffer solutions were made using BDH AnalaR-grade reagents and double-
distilled water.

Emulsion Preparation. Oil-in-water emulsions (20 wt % n-tetradecane,
0.5 wt % pure milk protein or 0.1 wt % sodium caseinate, 20 mM
imidazole buffer, pH 7) were prepared at room temperature using a
small-scale valve homogenizer (8) operating at a pressure of 300 bar.
Droplet-size distributions needed to calculate emulsion surface areas
were determined using a Coulter counter (TA II) with a 30 μm orifice
tube and 0.18 M NaCl as suspending electrolyte.

Emulsion Protein Exchange. An emulsion made with pure protein B was
washed free of unadsorbed protein by centrifuging at 10^4 g for 15
minutes at 20 °C, redispersing the cream in buffer, and then repeating
the procedure. The amount of unadsorbed protein in the aqueous phase
was determined by fast protein liquid chromatography (FPLC) using a
Pharmacia Mono-Q ion-exchange column with a linear salt gradient and a
UV detector with a fixed wavelength filter at 280 nm. To the washed
emulsion was added an equal volume of a buffered solution of protein
A. The mixed emulsion system (now 10 wt % n-tetradecane) was stirred
continuously at 20 °C for up to 48 hours. Aliquots removed during this
period were centrifuged, and the composition of the resulting aqueous
phase was determined by FPLC.

Surface Viscosity. Surface shear viscosity at the n-tetradecane—water
interface (pH 7, 25 °C) was measured using a Couette-type torsion-wire
surface rheometer (9,10) operating at a steady slow rotation speed of
1.3×10^{-3} rad s^{-1}. An adsorbed film of protein B, which had been aged
for several hours (typically 24 h), was exposed to protein A by
introducing the latter into the subphase below the interface.

Interfacial Tension. Measurements were made at the oil—water
interface (pH 7, 25 °C) using a Wilhelmy-plate torsion balance with a

5 cm wide mica plate (11). Quoted equilibrium tensions refer to the steady-state values reached after 24 hours.

Protein/Protein Competitive Adsorption

The α_{s1}-Casein/β-Casein System. These two proteins comprise over 75% of the casein in milk. Compared with most other food proteins, both are highly disordered and substantially hydrophobic (12). Of the two, β-casein is the more surface-active: it gives a lower steady-state tension at the oil—water interface (13), and it predominates at the interface in emulsions made with a mixture of the two caseins (14). Exchange experiments indicate that β-casein rapidly displaces α_{s1}-casein from the emulsion droplet surface, as shown in Figure 1 for the case of a 1:1 α_{s1}-casein/β-casein ratio. The redistribution of the two proteins between interface and bulk phase is accomplished within a few minutes of adding β-casein to the α_{s1}-casein-coated droplets. With three consecutive exchanges using a 1:10 α_{s1}-casein/β-casein ratio, it was established (14) that all the original α_{s1}-casein could be displaced from the surface by β-casein. We also found (14) that α_{s1}-casein added to β-casein-coated droplets will displace some β-casein, but to a much lesser extent than the other way round. These results indicate that competitive adsorption of α_{s1}-casein + β-casein in emulsions is a rapid reversible process.

Surface viscosity measurements are sensitive to the age, the structure and the composition of adsorbed protein films (9,13,15-17). The two individual caseins give surface viscosities which are quite low but still substantially different. For films adsorbed at pH 7 from 10^{-3} wt% protein solutions, the values after 24 h are 2.0 mN m^{-1} s for α_{s1}-casein and 0.4 mN m^{-1} s for β-casein. Figure 2 shows what happens when β-casein is exposed to a 24-hour-old film of α_{s1}-casein. Over a further period of 24 h, there is a gradual fall in surface viscosity to a steady value of 0.65 mN m^{-1} s, which is the same value as that reached by α_{s1}-casein + β-casein (same concentrations) after 48 h. This behaviour is consistent with a steady reversible approach to an adsorbed film whose composition is predominantly, but not exclusively, β-casein. The much slower approach to equilibrium than in the emulsion exchange experiment is presumably due to the much lower protein concentrations (by a factor of 10^3) and the much larger distances for mass transport.

The β-Casein/β-Lactoglobulin System. Here we have a mixture of a disordered protein, β-casein, and a structured globular protein, β-lactoglobulin. The emulsion exchange results presented in Figure 3 show two main points of difference from the α_{s1}-casein/β-casein case: firstly, there is little displacement of either protein from the droplet surface by the other, and, secondly, what little exchange does take place has still not reached equilibrium after 48 h. Figure 3a indicates that β-lactoglobulin slowly adsorbs at the droplet surface with no displacement of β-casein. Figure 3b shows that added β-casein rapidly reaches a steady surface concentration, and that the globular protein is initially partially displaced, but then slowly readsorbs over an extended period of time.

Compared with β-casein, the surface viscosity of β-lactoglobulin is very high (1200 mN m^{-1} s for a film adsorbed from 10^{-3} wt% protein

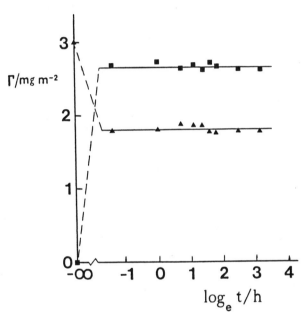

Figure 1. Protein exchange on mixing washed α_{s1}-casein emulsion with β-casein solution (1:1 protein ratio). Surface concentration Γ is plotted against logarithm of time t following mixing: ▲, α_{s1}-casein; ■, β-casein. (Reproduced with permission from reference 18. Copyright 1989 Oxford University Press.)

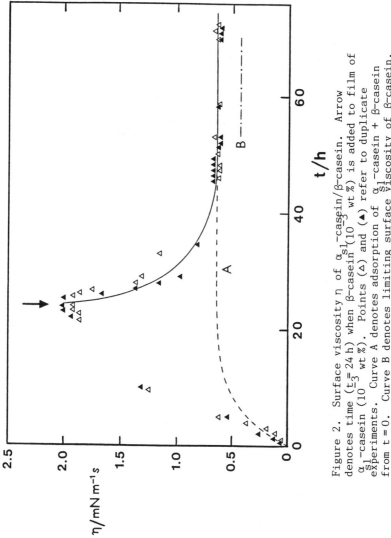

Figure 2. Surface viscosity η of α_{s1}-casein/β-casein. Arrow denotes time ($t = 24$ h) when β-casein (10^{-3} wt %) is added to film of α_{s1}-casein (10^{-3} wt %). Points ($\triangle$) and ($\blacktriangle$) refer to duplicate experiments. Curve A denotes adsorption of α_{s1}-casein + β-casein from $t = 0$. Curve B denotes limiting surface viscosity of β-casein. (Reproduced with permission from reference 28. **Copyright 1989 Royal Society of Chemistry.**)

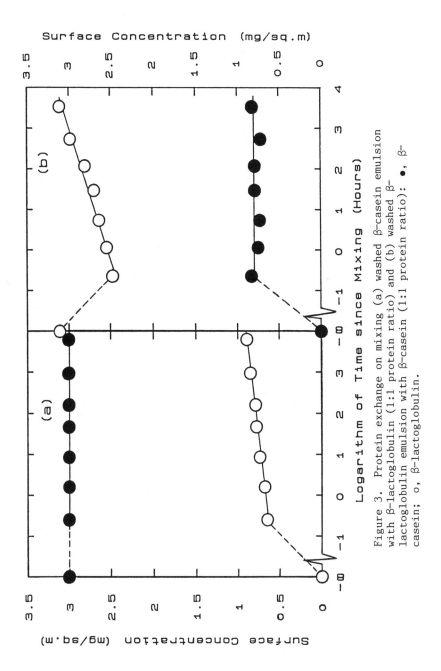

Figure 3. Protein exchange on mixing (a) washed β-casein emulsion with β-lactoglobulin (1:1 protein ratio) and (b) washed β-lactoglobulin emulsion with β-casein (1:1 protein ratio): •, β-casein; o, β-lactoglobulin.

solution after 24 h). The ability of β-lactoglobulin (10_3^{-3} wt %) to become incorporated into a 24-hour-old β-casein film (10^{-3} wt %) at the oil—water interface is illustrated in Figure 4. There is an increase in surface viscosity to a value of 400 mN m^{-1} s at 24 h after the addition of β-lactoglobulin into the subphase. This is a very high surface viscosity as compared with that for β-casein alone, but is appreciably lower than the value reached when both proteins are present together initially, and smaller still than the value reached in the absence of β-casein. Figure 5 shows the result of the converse experiment in which β-casein (10^{-3} wt %) is introduced into the subphase below a 24-hour-old film of adsorbed β-lactoglobulin. There is a drop of more than 50 % in the surface viscosity over a period of 2—3 h following the addition. But this is followed by a steady recovery to a high value of ca. 800 mN m^{-1} s at 24 h following the addition; this value is close to that at 48 h in Figure 4 for the two proteins adsorbing together. The overall behaviour is consistent with the increasing accumulation of β-lactoglobulin into the emulsion droplet surface layer as indicated in Figure 3b.

The β-Casein/α-Lactalbumin System. Here again we have a mixture of a disordered protein, β-casein, and a globular protein, α-lactalbumin. And again the exchange process is slow and effectively non-equilibrium in character. Addition of α-lactalbumin to β-casein-coated droplets gives behaviour similar to that shown in Figure 3a. We note, however, that α-lactalbumin is more readily displaced from the droplet surface by β-casein than is β-lactoglobulin. Figure 6 shows that the surface concentration of α-lactalbumin falls steadily with time, and that after 48 h over 60 % of the original α-lactalbumin has been displaced from the droplet surface. This contrasts sharply with Figure 3b which shows the surface concentration of β-lactoglobulin increasing with time after the initial drop.

Independent confirmation that it is easier to displace α-lactalbumin from the emulsion droplet surface than β-lactoglobulin is provided by recent electrophoretic mobility measurements (18) on whey protein emulsion droplets which were exposed to a solution of β-casein. It was found that β-casein addition has negligible effect on the mobility of β-lactoglobulin-coated droplets, implying that, while some β-casein may be adsorbed (see Figure 3a), β-lactoglobulin remains at the outer surface of the protein stabilizing layer. On the other hand, β-casein addition causes a substantial drop in the mobility of α-lactalbumin-coated droplets, implying replacement of some of the whey protein on the outer surface by β-casein.

The surface viscosity data in Figure 7 are also consistent with the displacement of α-lactalbumin from the oil—water interface by β-casein. On addition of β-casein (10^{-3} wt %) after 24 h, there is a sudden drop in surface viscosity from 500 to 50 mN m^{-1} s, with the value remaining essentially constant thereafter. This limiting value is similar to the low surface viscosity obtained after 48 h when the two proteins adsorb together, but it is still two orders of magnitude larger than that for pure β-casein, which shows that there is still a lot of α-lactalbumin present in the mixed adsorbed film.

The α-Lactalbumin/β-Lactoglobulin System. As with the binary mixtures of each whey protein with β-casein, this mixture of the two whey

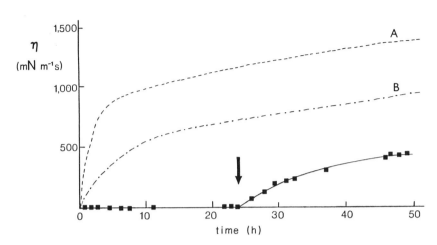

Figure 4.   Surface viscosity η of β-casein/β-lactoglobulin.   Arrow denotes time (t = 24 h) when β-lactoglobulin (10^{-3} wt %) is added to film of β-casein (10^{-3} wt %).   Curve A denotes adsorption of β-lactoglobulin from t = 0.   Curve B denotes adsorption of β-casein + β-lactoglobulin from t = 0.   (Reproduced with permission from reference 17. **Copyright 1990 International Journal of Biological Macromolecules.**)

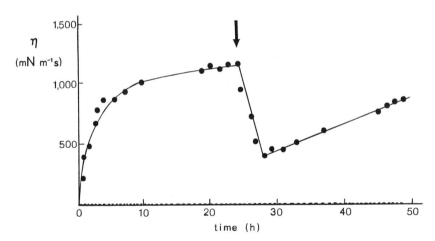

Figure 5.   Surface viscosity η of β-lactoglobulin/β-casein.   Arrow denotes time (t = 24 h) when β-casein (10^{-3} wt %) is added to film of β-lactoglobulin (10^{-3} wt %).   The dashed line at η ≈ 0 denotes adsorption of β-casein from t = 0.   (Reproduced with permission from reference 17. **Copyright 1990 International Journal of Biological Macromolecules.**)

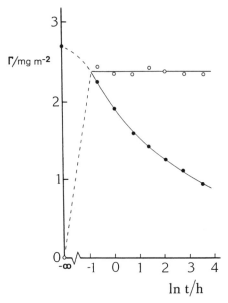

Figure 6.   Protein exchange on mixing washed α-lactalbumin emulsion with β-casein (1:1 protein ratio).   Surface concentration Γ is plotted against logarithm of time following mixing: ●, α-lactalbumin; o, β-casein.

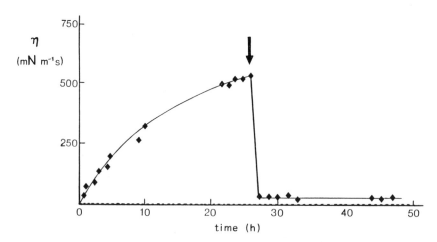

Figure 7.   Surface viscosity η of α-lactalbumin/β-casein.   Arrow denotes time ($t = 24$ h) when β-casein (10^{-3} wt %) is added to film of α-lactalbumin (10^{-3} wt %).   The dashed line at $\eta \simeq 0$ denotes adsorption of β-casein from $t = 0$.   (Reproduced with permission from reference 17. **Copyright 1990 International Journal of Biological Macromolecules.**)

proteins together also shows limited slow irreversible protein exchange between the bulk aqueous phase and the emulsion droplet surface. We have found ($\underline{18}$) that β-lactoglobulin will displace a significant amount of previously adsorbed α-lactalbumin from the droplet surface, but that α-lactalbumin will displace very little β-lactoglobulin. The fact that the latter is particularly difficult to displace may be associated with interfacial polymerization through disulphide bond interchange.

There is a substantial drop in surface viscosity on introducing β-lactoglobulin (10^{-3} wt %) into the subphase below a 24-hour-old film of α-lactalbumin (10^{-3} wt %) ($\underline{17}$). This is suggestive of some partial displacement of the latter by the former from the planar interface. In the reverse experiment, however, where α-lactalbumin is introduced into the subphase below an aged β-lactoglobulin film, there is no such change. On the contrary, as shown in Figure 8, the effect of adding the second globular protein is to "freeze in" the surface viscosity at the value of that for the pure β-lactoglobulin film at the time of the addition. It would appear that the presence of α-lactalbumin prevents the development of further intermolecular interactions between the β-lactoglobulin molecules at the oil—water interface. It has been suggested ($\underline{17}$) that, in the mixed whey protein system, there is an inhibition of the disulphide bond formation which is considered ($\underline{19}$) to be the origin of the strong time dependence of the surface rheology of pure β-lactoglobulin films.

Protein/Surfactant Competitive Adsorption

Small-molecule surfactants can affect the formation and stabilization of food emulsions in a number of ways. During homogenization, smaller droplets are produced in the presence of surfactants due to the more rapid reduction in interfacial tension than with protein alone ($\underline{20}$). Interaction of surfactant with protein may produce a thicker and stronger adsorbed layer ($\underline{21}$). Additional stabilization may be provided by liquid crystals at the interface ($\underline{22}$) or in the continuous phase between droplets ($\underline{23,24}$). Emulsions may be destabilized by an enhancement of globule clumping, caused by surfactant-induced partial disruption of protein interfacial layers during whipping or air incorporation, e.g., in the making of ice-cream ($\underline{25}$). Food-grade surfactants are often complex mixtures of undetermined composition, and so in this fundamental study of protein—surfactant competitive adsorption we prefer to use pure non-food-grade surfactants. Here, particular reference is made to octaoxyethylene glycol dodecyl ether ($C_{12}E_8$), which is a water-soluble non-ionic surfactant that has been widely studied ($\underline{26}$) as is available commercially in a highly purified form. $C_{12}E_8$ has an estimated HLB value of 13.1, and a critical micelle concentration (cmc) of 1.09×10^{-4} M ($\underline{11}$).

Figure 9 shows the equilibrium interfacial tension γ at 25 °C as a function of surfactant concentration c_s for $C_{12}E_8$ alone and for $C_{12}E_8$ + 0.1 wt % caseinate. The two sets of results become essentially coincident for $c_s \gtrsim 10^{-2}$ wt %. Beyond this concentration, the protein seems to be completely displaced from the oil—water interface, since the $C_{12}E_8$ excess concentration ($\Gamma = 1.27$ mg m^{-2}), as estimated from $d\gamma/d \ln c_s$ (below the cmc) using the Gibbs adsorption equation, is the same both in the presence and absence of 0.1 wt % caseinate.

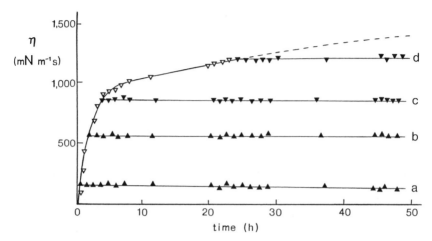

Figure 8. Surface viscosity η of β-lactoglobulin/α-lactalbumin:
▽, β-lactoglobulin (10^{-3} wt%) adsorbing from t = 0; ▲, ▼, α-
lactalbumin (10^{-3} wt%) added after (a) 5 min, (b) 30 min, (c) 4 h
and (d) 24 h. (Reproduced with permission from reference 17.
Copyright 1990 International Journal of Biological Macromolecules.)

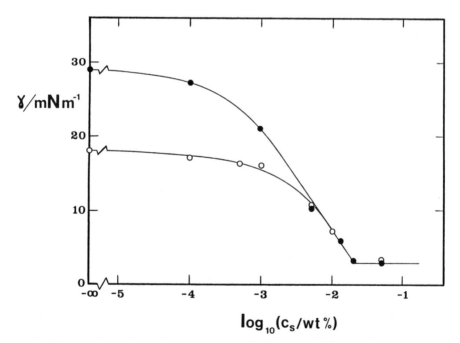

Figure 9. Interfacial tension γ of $C_{12}E_8$ + sodium caseinate as a
function of logarithm of surfactant concentration c_s: ●, no
protein present; o, 0.1 wt% caseinate present.

The much higher rate of lowering of the tension in the presence of surfactant leads to smaller average droplet sizes in caseinate-stabilized emulsions as indicated in Figure 10. In turn, these smaller droplets are more stable with respect to creaming, as shown by the extent of serum separation after 14 days storage; and they are more stable with respect to coalescence, as indicated by the extent of the change in droplet-size distribution over the same time period. Qualitatively similar trends of behaviour have also been found (20) with caseinate + the oil-soluble non-ionic surfactant $C_{12}E_2$.

A very low bulk phase concentration of $C_{12}E_8$ has a large effect on the protein film surface viscosity, as illustrated in Figure 11. In the absence of surfactant, a caseinate film (0.1 wt%) develops a surface viscosity of ca. 20 mN m^{-1} s after 24 h. However, the presence of just 10^{-4} wt% $C_{12}E_8$ prevents the formation of a film with any measurable surface viscosity. Introducing the same concentration of $C_{12}E_8$ after 24 h of protein adsorption gives the same effect. The tension data in Figure 9 would seem to indicate that caseinate is predominant at the interface at $c_s = 10^{-4}$ wt%. So, it would seem that the reason for the loss of surface viscosity at $c_s = 10^{-4}$ wt% is not complete displacement of protein from the interface, but rather a disruption of protein—protein interactions in the film, due to the lubricating action of adsorbed surfactant, which leads to a lower resistance to flow (11).

The behaviour of caseinate + $C_{12}E_8$ is probably typical of what one can expect for a system containing a mixture of a food protein and a non-ionic (non-interacting) water-soluble surfactant. On the other hand, the situation is more complicated with anionic surfactants which interact strongly with proteins, both in bulk solution and at the interface (27). This is illustrated in Figure 12 by interfacial tension data for sodium dodecyl sulphate (SDS) + 0.1 wt% caseinate at 25 °C. At the highest surfactant concentration (well above the cmc), the tension for SDS + caseinate is the same as that for SDS alone, suggesting that under these conditions the protein is completely displaced. At lower SDS concentrations, however, the situation is more complex. At surfactant concentrations just above the cmc ($c_s \simeq$ 0.1 wt%), the tension in the presence of protein is significantly higher than that for SDS alone, and there is a plateau region in the SDS + caseinate data corresponding to an effectively constant tension over two orders of magnitude of SDS concentration (down to $c_s \simeq 10^{-3}$ wt%). Presumably, in the plateau region, it is more favourable in terms of free energy for surfactant to bind cooperatively onto protein than it is for surfactant to displace protein from the interface. At the end of the plateau region, when there are no more binding sites available on the protein, free SDS molecules begin to displace the protein—surfactant complexes from the interface. Similar trends of behaviour occur also with SDS + gelatin (27).

There are two main differences between the non-ionic (i.e. non-interacting) surfactant (e.g. $C_{12}E_8$) and the ionic (interacting) surfactant (SDS). With the ionic surfactant, at concentrations below that required for protein displacement, the formation of a protein—surfactant electrostatic complex at the interface may lead to a substantial increase in surface viscosity as compared with the pure protein film. The second point is that, because it binds to the protein, not all the ionic surfactant is available for adsorption at

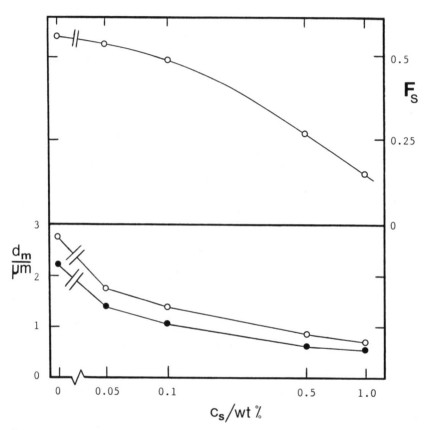

Figure 10.   Effect of $C_{12}E_8$ on stability of emulsions (0.1 wt %
caseinate, 20 wt % n-tetradecane, pH 7, 25 °C).   Lower graph shows
median droplet diameter d_m against surfactant concentration c_s:
•, fresh emulsion; o, after 14 days storage.   Upper graph shows
corresponding fraction F_s of each emulsion sample which exists as
serum layer after 14 days storage.

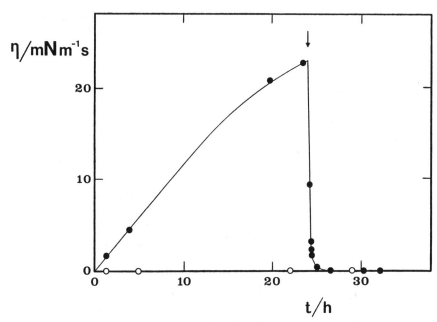

Figure 11. Surface viscosity η of caseinate/$C_{12}E_8$. Arrow denotes time (t = 24 h) when $C_{12}E_8$ (10^{-4} wt %) is added to film of caseinate (10^{-1} wt %). Open circles denote adsorption of caseinate + $C_{12}E_8$ from t = 0.

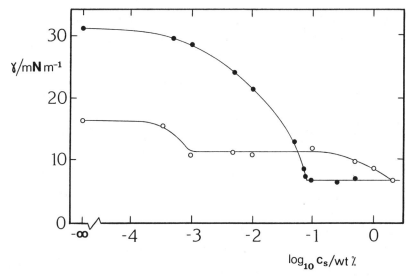

Figure 12. Interfacial tension γ of SDS + sodium caseinate as a function of logarithm of surfactant concentration c_s: •, no protein present; o, 0.1 wt % caseinate present.

the oil—water interface. This means that a higher total concentration of SDS is required ultimately to displace the protein than is required by the non-ionic surfactant.

Acknowledgment

Continued financial support for this research by the Agricultural and Food Research Council (U.K.) is gratefully acknowledged.

Literature Cited

1. Dickinson, E.; Stainsby, G. Colloids in Food; Applied Science: London, 1982.
2. Dickinson, E. In Food Structure—Its Creation and Function; Blanshard, J. M. V.; Mitchell, J. R., Eds.; Butterworths: London, 1988; pp. 41-57.
3. Dickinson, E. In Gums and Stabilisers for the Food Industry; Phillips, G. O.; Wedlock, D. J.; Williams, P. A., Eds.; IRL Press: Oxford, 1988; Vol. 4, pp. 249-263.
4. Dickinson, E. Colloids Surf. 1989, 42, 191-204.
5. Dickinson, E.; Euston, S. R. In Food Polymers, Gels and Colloids; Dickinson, E., Ed.; Royal Society of Chemistry: London; in press.
6. Zittle, C. A.; Custer, J. H. J. Dairy Sci. 1963, 46, 1183-1188.
7. Annan, W. D.; Manson, W. J. Dairy Res. 1969, 36, 259-268.
8. Dickinson, E.; Murray, A.; Murray, B. S.; Stainsby, G. In Food Emulsions and Foams; Dickinson, E., Ed.; Royal Society of Chemistry: London, 1987; pp. 86-99.
9. Dickinson, E.; Murray, B. S.; Stainsby, G. J. Colloid Interface Sci. 1985, 106, 259-262.
10. Euston, S. E. Ph. D. thesis, University of Leeds, 1989.
11. Woskett, C. M. Ph. D. thesis, University of Leeds, 1989.
12. Swaisgood, H. E. In Developments in Dairy Chemistry; Fox, P. F., Ed.; Elsevier Applied Science: London, 1982; Vol. 1, pp. 1-59.
13. Castle, J.; Dickinson, E.; Murray, B. S.; Stainsby, G. In Proteins at Interfaces; Brash, J. L.; Horbett, T. A., Eds.; ACS Symposium Series No. 343; American Chemical Society: Washington, DC, 1987; pp. 118-134.
14. Dickinson, E.; Rolfe, S. E.; Dalgleish, D. G. Food Hydrocolloids 1988, 2, 397-405.
15. Boyd, J. V.; Mitchell, J. R.; Irons, L.; Mussellwhite, P. R.; Sherman, P. J. Colloid Interface Sci. 1973, 45, 478-486.
16. Graham, D. E.; Phillips, M. C. J. Colloid Interface Sci. 1980, 76, 240-250.
17. Dickinson, E.; Rolfe, S. E.; Dalgleish, D. G. Int. J. Biol. Macromol. 1990, 12, 189-194.
18. Dickinson, E.; Rolfe, S. E.; Dalgleish, D. G. Food Hydrocolloids 1989, 3, 193-203.
19. Dickinson, E.; Murray, B. S.; Stainsby, G. In Advances in Food Emulsions and Foams; Dickinson, E.; Stainsby, G., Eds.; Elsevier Applied Science: London, 1988; pp. 123-162.
20. Dickinson, E.; Mauffret, A.; Rolfe, S. E.; Woskett, C. M. J. Soc. Dairy Technol. 1989, 42, 18-22.
21. Wüstneck, R.; Müller, H.-J. Colloid Polym. Sci. 1986, 264, 97-102.

22. Friberg, S.; Jansson, P. O.; Cederberg, E. J. Colloid Interface Sci. 1976, 55, 614-623.
23. Barry, B. W.; Saunders, G. M. J. Colloid Interface Sci. 1972, 41, 331-342.
24. Dickinson, E.; Narhan, S. K.; Stainsby, G. J. Food Sci. 1989, 54, 77-81.
25. Madden, J. K. In Foams: Physics, Chemistry and Structure; Wilson, A. J., Ed.; Springer-Verlag: London, 1989; pp. 185-196.
26. Mitchell, D. J.; Tiddy, G. J. T.; Waring, L.; Bostock, T.; MacDonald, M. P. J. Chem. Soc., Faraday Trans. 1 1983, 79, 975-1000.
27. Dickinson, E.; Woskett, C. M. In Food Colloids; Bee, R. D.; Richmond, P.; Mingins, J., Eds,; Royal Society of Chemistry: London, 1989; pp. 74-96.
28. Dickinson, E.; Rolfe, S. E.; Dalgleish, D. G. In Food Colloids; Bee, R. D.; Richmond, P.; Mingins, J., Eds.; Royal Society of Chemistry: London, 1989; pp. 377-381.

RECEIVED August 16, 1990

Chapter 10

Protein–Glyceride Interaction
Influence on Emulsion Properties

A. Martinez-Mendoza and P. Sherman

Department of Food and Nutritional Sciences, King's College, University of
London, Kensington Campus, Campden Hill Road, London W8 7AH,
United Kingdom

Partial replacement of water soluble meat proteins (WSMP) by
glyceryl monostearate and glyceryl distearate in concentrated corn
oil-in-water emulsions influences their rheological properties, the
rate of drop coalescence, interfacial tension reduction with time
and the rheological properties of the emulsifier film around the
oil drops. The pH also exerts an effect. The
WSMP/monoglyceride/diglyceride ratio is critical, with maximum
influence exerted when using a WSMP/glycerides ratio of 1/1
(monoglyceride/ diglyceride ratio 3/1) near to the WSMP's isoelectric point. The monoglyceride and diglyceride adsorb at the oil-
water interface before the WSMP. They enter into some form of
association and interact with WSMP which is adsorbed later.
Association of the two glycerides increases the interfacial area
available to the WSMP for adsorption. This is also facilitated
near to the WSMP's isoelectric point when the protein molecules
are in a more compact configuration. Surface pressure increases
with increasing WSMP adsorption and this leads to the projection
of protein hydrophilic groups from drop surfaces. Interlinking of
the loops on neighboring drops, and loop concentration, are primarily responsible for the rheological properties of the emulsions.

In recent years increasing consideration has been given to the use of dairy and
vegetable proteins for the improvement of existing food products and the
development of new ones.

Many foods prepared from alternative protein sources take the form of oil-
in-water emulsions, e.g., filled milks, acid protein beverages, emulsified wheat glu-
ten for baked products and pasta, and simulated meat emulsions. The surface
chemistry of some food proteins has been studied in some detail. Although the
information derived is of value, the fact remains that food formulations often
include non-protein surface active components. Commercial grade glyceryl
monostearate, which is one such example, contains a significant proportion of glyceryl distearate.

0097–6156/91/0448–0130$06.00/0
© 1991 American Chemical Society

Concentrated oil-in-water emulsions stabilized with sodium caseinate (*1,2,3*), or chemically modified 7S soy protein (*4,5,6*), in conjunction with glyceryl monostearate and glyceryl distearate showed maximum values of their viscoelasticity parameters, minimum rate of drop coalescence and maximum values of the interfacial viscoelasticity parameters at a critical protein/monoglyceride/diglyceride ratio at each pH investigated. The critical ratio differed for the two proteins and the greatest effect was exerted at a pH near to the isoelectric point of either protein.

The previous studies have been extended to interactions between water soluble meat proteins (WSMP), monoglyceride and diglyceride in concentrated corn oil-in-water emulsions. WSMP promote the stability of the meat emulsion in ground meat products. During heat processing these products become unstable, and coalescence of fat particles leads to the development of undesireable fat pockets. The aim of this work was to ascertain whether the incorporation of glyceryl monostearate and glyceryl distearate would reduce this tendency. A past empirical study of meat emulsion stability (*7*) suggested that the addition of 0.1-0.3% (wt/wt) monoglyceride and diglyceride increased fat release.

Materials and Methods

Materials. Pure corn oil (CPC UK Ltd., density 0.919 gm/cm at 20°C) and double distilled water were used.

The glyceryl monostearate contained at least 90% monoglyceride. The glyceryl distearate had a glyceride content of at least 80%.

WSMP were extracted from fresh pork fillets (*8*), freeze dried, and stored at 1°C in glass jars until required.

Concentrated oil-in-water emulsions with an oil phase/water phase ratio of 60/40 (wt/wt) were prepared as described elsewhere (*9*). Homogenization was carried out with a Rannie pressure homogenizer at 3.5 MPa. Four series of emulsions were prepared. One series contained 2.0% WSMP. In the other three series 0.5%, 1.0% and 1.5% WSMP were replaced by an equivalent weight of monoglyceride plus diglyceride. The monoglyceride/diglyceride ratios used in these three series of emulsions were 1/3, 1/1, and 3/1 at pHs 3.5, 5.5 and 7.5.

3.0M HCl or 3.0M NaOH was used to adjust the pH. Following preparation all emulsions were stored in a refrigerator at 4°C.

For both interfacial tension-time studies and the rheological measurements at the corn oil-water interface, four series of tests were made. In one series 1.0% (wt/wt) WSMP were used, while in the other three series 0.25%, 0.50% and 0.75% (wt/wt) were replaced by an equivalent weight of monoglyceride plus diglyceride, respectively. Monoglyceride/diglyceride ratios of 1/3, 1/1 and 3/1 were used in each of the latter three series. In all four series measurements were made at pHs 2.5, 3.5 and 9.5.

Creep Compliance—Time Response of Corn Oil-In-Water Emulsions. A Deer rheometer Mark 11 (Rheometer Marketing Ltd., Leeds, England) was used with cone and plate attachments. The cone angle was 2° and its diameter was 6.5 cm. The plate diameter was 8.0 cm. All measurements were made at 25.0 + 0.1°C at a low constant stress within the linear viscoelastic region.

Size Distribution and Electrophoretic Mobility of Oil Drops in Corn Oil-In-Water Emulsions. The drop size distribution in each emulsion was determined periodically with a Joyce Loebl photosedimentometer Mark 111 (*10*). The slow rate of drop coalescence was derived from the rate of increase in the mean volume diameter with time at 4 ° C (*11*).

A microelectrophoresis unit (Rank Bros., Bottisham, Cambridge, England) was used for electrophoretic mobility measurements.

Time-Dependent Interfacial Tension Reduction at the Corn Oil-Water Interface. A Kruss Digital Tensiometer K1O (Kruss GmbH, Hamburg, West Germany) fitted with a Wilhelmy plate monitored the reduction of interfacial tension with time.

WSMP was introduced into the aqueous phase and the pH was adjusted with 3.0 M HCl or 3.0 M NaOH. Monoglyceride and diglyceride were introduced into the oil phase. Both phases were heated to $45.0\,^{\circ}C + 0.1\,^{\circ}C$ and the oil phase was layered on the aqueous phase. The temperature of $45.0\,^{\circ}C + 0.1\,^{\circ}C$ was maintained throughout all measurements.

Rheological Properties at the Corn Oil-Water Interface. Interfaces were developed, and maintained, as for interfacial tension-time studies. The viscoelastic properties were examined with an oscillating ring rheometer (*12*). This rheometer applies a sinusoidally varying torque to a platinum ring located at the oil-water interface so that the interface is subjected to a shear stress.

Interfacial elasticity ($G_s{'}$) is given by

$$G_s{'} = (I_0/G_f)(W^2 - W_0^2) \tag{1}$$

where

$$G_f = 4\pi(1/R_1^2 - 1/R_2^2) \tag{2}$$

I_0 is the moment of inertia of the ring rheometer's moving parts, W and W_0 are the working pulsatances (or frequencies) at the interface, G_f is the geometry factor, and R_1 and R_2 are the respective radii of the platinum ring and the glass dish in which the corn oil–water interface is developed.

Results

Rheological Properties of Corn Oil-In-Water Emulsions. All the emulsions, irrespective of protein, monoglyceride and diglyceride content, or pH, exhibited a creep compliance-time response which could be represented (*9*) by

$$J(t) = J_0 + J_1[1 - \exp(-t/\tau_1)] + J_2[1 - \exp(-t/\tau_2)] + t/\eta_N \tag{3}$$

where J_0 ($= 1/E_0$) is the instantaneous elastic compliance, J_1 ($= 1/E_1$,) and J_2 ($= 1/E_2$) are the first and second retarded elastic compliances, τ_1 and τ_2 are the first and second retardation times, ν_N is the Newtonian viscosity, E_0 is the instantaneous elastic modulus, and E_1 and E_2 are the first and second retarded elastic moduli.

At any storage time and WSMP/monoglyceride(M)/ diglyceride(D) ratio the highest values of E_0, E_1, E_2, and ν_N were obtained at pH 5.5 (9). Table I summarizes the influence of this ratio on E_0 at pH 5.5 for selected storage times, with G representing monoglyceride plus diglyceride. Only E_0 values are quoted because they relate to the strength of the interlinked drop aggregates before they begin to break down. E_1, E_2, and ν_N showed similar trends.

E_0 increased at all three WSMP/M/D ratios during the first five days storage to maximum values even though the mean drop diameter increased continuously. When drop coalescence is dominant during storage of emulsions, E_0 decreases. In the present study E_0 began to decrease after five days. The initial, maximum and final values of E_0 were influenced by WSMP/M/D ratio with the greatest effect exerted when using 1.0% WSMP in conjunction with 1.0% G (M/D ratio 3/1). At this latter ratio E_0 was significantly higher than when 2.0% WSMP was used.

Table I.   E Data for Corn Oil-in-Water Emulsions at pH 5.5.

M/D Ratio	Days Storage	$E_0(Nm^{-2}X10^{-3})$			
		2.0% WSMP +0% G	1.5%WSMP +0.5% G	1.0%WSMP +1.0% G	0.5%WSMP +1.5% G
—	0.12	2.01	—	—	—
	5	15.81	—	—	—
	15	3.08	—	—	—
	0.12	—	10.26	17.12	3.49
1/3	5	—	14.68	24.93	8.37
	15	—	7.48		
	0.12	—	11.02	37.87	4.73
1/1	5	—	19.76	43.13	9.46
	15	—	9.80	8.14	1.61
	0.12	—	20.50	38.21	6.39
3/1	5	—	34.45	46.70	13.65
	15	—	12.77	20.21	3.42

Rate of Drop Coalescence. The initial mean volume diameters were 0.21-0.2 μm, and after 15 days 0.26-0.38 μm. The slow rate of drop coalescence (Table II) was influenced by WSMP/M/G ratio and by pH, the lowest rate occurring at pH 5.5 with 1.0% WSMP plus 1.0% G (M/D ratio 3/1). Under these conditions the rate was significantly lower than for emulsions incorporating 2.0% WSMP.

Electrophoretic Mobility of Oil Drops. Zero mobility was found at pH 5.5 irrespective of WSMP/M/D ratio (Table III). The mobility was higher at pH 7.5 than at pH 3.5 at each of the three ratios.

At a constant pH of 7.5 or 3.5 the mobility decreased as the M/D ratio increased from 1/3 to /1 at a constant WSMP/M/D ratio.

Interfacial Tension-Time Reduction. Irrespective of WSMP/M/D ratio the interfacial tension decreased for about 60 min. A steady state value (γ_{ss}) was achieved at this time. The interfacial tension between pure corn oil and distilled water was 18.4 mNm.

Table II. E Data for Corn Oil-in-Water Emulsions

M/D Ratio	pH	Slow Rate of Coalescence $(X10^7 \ sec^{-1})$			
		2.0% WSMP +0% G	1.5%WSMP +0.5% G	1.0%WSMP +1.0% G	0.5%WSMP +1.5% G
—	7.5	5.79	—	—	—
	5.5	3.26	—	—	—
	3.5	4.67	—	—	—
	7.5	—	4.40	5.95	4.67
1/3	5.5	—	2.52	1/59	3.06
	3.5	—	4.47	1.85	5.14
	7.5	—	5.15	6.93	4.99
1/1	5.5	—	3.58	2.32	3.09
	3.5	—	4.51	4.24	5.17
	7.5	—	4.00	2.25	4.12
3/1	5.5	—	2.28	0.73	2.30
	3.5	—	4.27	1.63	2.81

Table III. Electrophoretic Mobility of Oil Drops in Corn
Oil-in-Water Emulsions

M/D Ratio	pH	Electrophoretic Mobility $(m \ sec^{-1}$ per volt cm-1)			
		2.0% WSMP +0% G	1.5%WSMP +0.5% G	1.0%WSMP +1.0% G	0.5%WSMP +1.5% G
—	7.5	6.05	—	—	—
	5.5	—	—	—	—
	3.5	1.59	—	—	—
	7.5	—	6.14	6.93	8.62
1/3	5.5	—	—	—	—
	3.5	—	2.09	2.15	2.37
	7.5	—	6.06	6.77	7.0
1/1	5.5	—	—	—	—
	3.5	—	1.83	1.98	2.20
	7.5	—	5.58	5.64	6.14
3/1	5.5	—	—	—	—
	3.5	—	1.49	1.57	1.83

Both WSMP/M/D ratio and pH influenced γ_{ss} (Table IV), with the minimum value at 0.5% WSMP/0.5% G (M/D ratio 3/1) at pH 3.5. This value was much lower than that for 1.0% WSMP. At constant WSMP/G ratio and pH, γ_{ss} decreased as the M/D ratio increased from 1/3 to 3/1. Increasing the pH from 2.5 to 3.5 at constant WSMP/M/D ratio decreased γ_{22}, and then it increased with further rise of pH from 3.5 to 9.5

Table IV. Influence of WSMP/M/D Ratio on γ_{ss}

		γ_{ss} (mNm-1)			
M/D Ratio	pH	1.0% WSMP +0% G	0.75%WSMP +0.25% G	0.5%WSMP +0.5% G	0.25%WSMP +0.75% G
	9.5	12.3	—	—	—
—	3.5	7.0	—	—	—
	2.5	6.2	—	—	—
	9.5	—	5.0	7.4	6.8
1/3	3.5	—	2.8	3.2	5.4
	2.5	—	3.9	7.2	5.8
	9.5	—	4.7	4.6	5.5
1/1	3.5	—	2.8	2.6	3.8
	2.5	—	3.1	3.0	4.2
	9.5	—	3.0	1.8	3.7
3/1	3.5	—	2.5	0.6	2.3
	2.5	—	2.8	1.6	3.5

Interfacial Rheological Properties. All interfacial films exhibited viscoelasticity irrespective of the presence or absence of M and D. $G_s{}'$ increased over several hours and achieved steady state values ($G_{ss}{}'$) after about 6 hours. Table V summarizes the influence of WSMP/M/D ratio and pH on values $G_{ss}{}'$. At each pH the highest value was observed when using 0.5% WSMP/0.5% G_N (M/D ratio 3/1). With this ratio the highest value of $G_{ss}{}'$ was obtained at pH 3.5.

Discussion

The data obtained by all of the experimental techniques utilized confirm the improvement achieved when WSMP is partially replaced by a blend of M and D, and that the WSMP/M/D ratio is important. These conclusions support previous observations regarding the critical nature of the protein/M/D ratio when using sodium caseinate (*1,2,3*) or modified 7S soy protein (*4,5,6*) in conjunction with M and D, although the optimum ratio depends on the protein used.

The main constituents of WSMP are myogen and myoglobin which have molecular weight ranges of 30,000 - 100,000 and 16,890 - 18,000, respectively (*13*). Myoglobin is a globular protein with a 70% α helical chain and a heme residue. Myogen is the term used to describe a heterogeneous mixture of at least 100 - 200 proteins.

The constitution of the WSMP-M-D film adsorbed on the surface of the oil drops can be deduced by reasoning as follows. Because of their relatively low molecular weights M and D should diffuse to the oil-water interface, and be

Table V. Influence of WSMP/M/D Ratio on G_{ss}'

M/D Ratio	pH	G_{ss}' (mNm^{-1})			
		1.0% WSMP +0% G	0.75%WSMP +0.25% G	0.5%WSMP +0.5% G	0.25%WSMP +0.75% G
—	9.5	1.6	—	—	—
	3.5	26.0	—	—	—
	2.5	1.6	—	—	—
	9.5	—	1.3	1.6	0.9
1/3	3.5	—	22.3	26.0	25.1
	2.5	—	20.7	18.4	18.0
	9.5	—	1.4	1.8	1.1
1/1	3.5	—	26.5	27.6	26.0
	2.5	—	21.5	23.2	20.0
	9.5	—	1.8	2.0	1.4
3/1	3.5	—	30.3	36.0	33.3
	2.5	—	24.6	28.1	20.3

adsorbed more quickly than WSMP. When the M/D ratio in the oil phase prior to emulsification is 3/1 the M and D enter into some form of association after adsorprtion (14). At other M/D ratios the interface contains a mixture of associated M - D molecules plus excess M or D. The concentration of M and D molecules associated at the interface increases as the WSMP/G ratio decreases when using a constant M/D ratio of 3/1. Association of adsorbed M and D molecules results in their occupying a smaller area of interface than if they did not associate, so that a larger area of interface is available for the slower diffusing and adsorbing WSMP molecules. Many globular proteins unfold partially at the oil-water interface. The pH influences the configuration of the WSMP molecules. They have a more compact form at, or near, their isoelectric point and then diffuse to, and adsorb at, the interface more readily. Thus, both the number of associated M and D molecules and pH influence the degree to which WSMP can enter the interfacial region. It might be argued that WSMP adsorbs on to a previously adsorbed layer of M and D molecules, but the interfacial rheological data do not support this view. The WSMP/M/D ratio would not exert such an influence on G_{ss}' in this case. A similar model to the one proposed here as been suggested (15) for the interaction between sodium caseinate and M at the oil-water interface.

As the interfacial tension decreases, and the surface pressure increases, the adsorbed WSMP molecules alter their configuration. Hydrophobic regions remain at the interface while hydrophilic groups project outwards into the thin films of aqueous phase between adjacent oil drops. It is the hydrophobic regions which interact with the previously associated M and D molecules. Optimum interaction occurs near to the WSMP's isoelectric point when using a 1/1 WSMP/G ratio with an M/D ratio of 3/1. Under these conditions the maximum concentration of WSMP is adsorbed and generates the largest number of hydrophilic loops per unit area of interface. Interlinking of these loops is primarily responsible for the magnitude of G_{ss}', and the values are much higher than if van der Waals' attraction between neighboring drops was the only operative factor.

One final point. Detailed analysis of both the interfacial tension and interfacial rheology data (*16*) supports the view that adsorption of WSMP is a two-stage process, as for other proteins. This belief has been questioned (*17*) with the argument that the inflexion point in $d\pi/dt$ vs. plots, where π is the surface pressure, is an artifact arising from the invalidity of the assumption that dn/dt is constant in the equation

$$(dn/dt)_{\pi} = (d\pi/dt)(dn/d\pi) \tag{4}$$

where n is the surface concentration of protein. However, when interfacial rheological data were plotted as $1/G_s{}'$, vs. $1/t$ an inflexion point was observed between two linear regions with different gradients, the larger gradient corresponding to higher $1/t$ values. This indicates that it is the second stage of the adsorption process, involving configurational changes, hydrophilic loop formation and interlinking of loops on neighboring oil drops, which primarily influences $G_s{}''$.

Acknowledgment

The authors gratefully acknowledge the assistance received from Grinsted Products Research Laboratories, Brabrand, Denmark, by their provision of glyceryl monostearate and glyceryl distearate, and from Snow Brand Milk Products Technical Research Institute, Kawagoe, Saitama, Japan in providing glyceryl monostearate. These were the glycerides used in the present study.

Literature Cited

1. Doxastakis, G.; Sherman, P. In *Instrumental Analysis of Foods. Recent Progress* Vol. 2. Charalambous, G.; Inglett, G., Eds; Academic Press, New York, 1983; pp. 219-35.
2. Doxastakis, G.; Sherman, P. *Colloid Polymer Sci.* 1984, *262*, 902.
3. Doxastakis, G.; Sherman, P. *Colloid Polymer Sci.* 1986, *264*, 254.
4. Reeve, M.J.; Sherman,P. *Food Microstructure* 1986, **5**,163.
5. Reeve, M.J.; Sherman, P. *Colloid Polymer Sci.* 1988. *266*, 930.
6. Reeve, M.J.; Sherman, P. J. *Dispersion Sci. Technol.* 1988, **90**, 343.
7. Mayer, J.A.; Brown, W.L.; Giltner, N.E.; Guin, J.R. *Food Technol.* 1964,**18** 1796.
8. Saffle, R.L.; Galbreath, J.W. *Food Technol.* 1964, **18**, 1943.
9. Martinez-Mendoza, A.; Sherman, P. *J. Dispersion Sci. Technol.* 1988/1989, **9**, 537.
10. Sherman, P.; Benton,M. J. *Texture Studies*, 1980, **11**, 1.
11. Vernon Carter, E.J.; Sherman, P. *J. Texture Studies*, 1980, **11**, 351.
12. Sherrif, M.; Warburton, B. In *Theoretical Rheology* Hutton, J; Pearson, J.R.A.; Walters, K.,Eds; *Applied Science*, London, 1975; pp. 299-308.
13. Ockerman, H. In *Food Colloids* Graham,H.D. Ed; Avi, Westport, U,S.A., 1977, p. 240.
14. Kako, M.; Kondo, S. *J. Colloid Interface Sci.* 1979, **6**, 163.
15. Paquin, P.; Britten, M,; Laliberte, M.F.; Boulet, M. In *Proteins At Interfaces. Physicochemical and Biochemical Studies* Brash, J.L.; Horbett, T.A. Eds; ACS, Washington,D.C., pp. 677-686.
16. Martinez-Mendoza, A.; Sherman, P. *J.Dispersion Sci. Technol.* to be published.
17. MacRitchie, F.; *Colloids and Surfaces* 1989, **41**, 25.

RECEIVED September 21, 1990

Chapter 11

Thermodynamics of Interfacial Films in Food Emulsions

N. Krog

Grindsted Products, Edwin Rahrsuej 38, 8220 Brabrand, Denmark

Interfacial tension of oil/water systems
containing emulsifiers has been studied as a
function of decreasing temperature in the range
from 40°C to 5°C. With 0.2% saturated
monoglycerides in the oil phase, the interfacial
tension decreases from 18 mN/m at 40°C to 2-4
mN/m at 5°C. The interfacial tension decreases
slowly in the upper temperature range until a
critical temperature (T_χ) is reached. Below
T_χ the interfacial tension decreases fast to
a minimum below 5 mN/m. The temperature-induced
changes vary with fatty acid chain length, degree
of unsaturation and hydrophilic properties of the
emulsifiers. The process is reversible.
Interfacial films of milk proteins are not
affected by temperature variations. When
monoglycerides are present in the oil phase,
proteins are desorbed from the interface at
temperatures below T_χ of the monoglyceride.
The interfacial exchange of protein and
monoglycerides at low temperatures correlates
well with a reduction in protein load on fat
globules in whippable emulsions.

Many food emulsions are stabilized by a layer of protein adsorbed
at the surface of the fat globules. This is typical of emulsions
such as homogenized milk, cream and ice cream mix before
freezing. During ice cream production the mix is whipped and
frozen, simultaneously forming an aerated frozen emulsion in
which the air cells are covered by a layer of fat globules and
fat crystals from disrupted fat globules.

0097–6156/91/0448–0138$06.00/0
© 1991 American Chemical Society

The orientation of fat globules at the air/serum interface is promoted by the desorption of the protein layer surrounding the fat globules, which increases the hydrophobicity of the fat globule surfaces. This desorption takes place during the aging of the mix at low temperature before freezing. Surface active lipids (emulsifiers) can adsorb at the fat globule surface in preference to the proteins during the homogenization of the emulsion, or replace the protein during the aging process.

Both lipophilic emulsifiers such as monoglycerides and the much more hydrophilic emulsifiers such as polysorbates can be used in ice cream. It has been shown that polysorbate reduces the amount of adsorbed protein at the fat globule surface (referred to as protein load) at the early stage of the formation of the emulsion (1). Monoglycerides, however, do not affect the initial protein load on fat globules in ice cream mix as much as polysorbates. Consequently, the function of monoglycerides still needs to be studied.

It has been found that the protein load on the fat globules in mix is reduced during the aging period at 5°C, and monoglycerides enhance this reduction. The interfacial tension of oil/water model systems which contain milk proteins in the water phase and monoglycerides in the oil phase decreases when the temperature is lowered from 40°C to 5°C, and this behavior correlates well with the desorption of protein from fat globule surfaces (2). The interfacial behavior of monoglyceride films has been studied in more detail, and the results of this work are described in the present work.

Methods and Materials

The interfacial tension of oil/water systems was measured on a Kruss Tensiometer, model 10, using the Wilhelmy plate method. The interfacial tension (γ) was recorded continuously by connecting a two-channel recorder to the tensiometer. The second channel on the recorder was used to monitor the temperature of the oil/water system in the tensiometer via a resistance thermometer. The temperature of the oil/water phases was controlled by a programmable thermostat, which allowed the temperature to be changed from 40°C to 5°C or lower at a rate of 0.3°C/min. Measurements were started at 40°C after preheating the oil phase and the water phase to 40°C separately. Initially, the interfacial tension was measured at 40°C for 1-2 hours to obtain a state of equilibrium between the oil and water phases. Then the temperature was decreased to 5°C at 0.3°C/min and kept for one hour at 5°C, after this period at 5°C the temperature was increased again to 40°C, at the maximum heating rate of 2°C/min. Finally, the temperature was kept at 40°C for 1-2 hours to re-establish equilibrium at 40°C.

High oleic sunflower oil was used as oil phase (Trisun 80 made by SVO Enterprises, Eastlake, Ohio). The monoglycerides were commercially distilled with a minimum total monoglyceride content of 95%. Saturated distilled monoglycerides (GMS) were made from hydrogenated soybean oil. Mono-unsaturated distilled mono-glycerides (GMO) were made from partially hydrogenated soy bean oil, iodine value 60, containing approx. 70 % oleic acid, 5% linoleic acid and approx. 25% saturated fatty acids. Polyunsaturated, distilled monoglycerides (GML) were made from sunflower oil, containing 69% linoleic acid, 20% oleic acid and 11% saturated fatty acids. Monoglycerides based on pure fatty acids such as glycerolmonomyristin (GMM), glycerolmonopalmitate (GMP) and glycerolmonobehenate (GMB) were made by esterification and subsequent distillation. The diacetyl tartaric acid esters of monoglycerides were commercial products (PANODAN AM referred to as DATEM I, and PANODAN 580 referred to as DATEM III. All emulsifiers used were made by Grindsted Products, Denmark. Pure, distilled water or solutions of 0.25% non-fat skim milk in distilled water was used as water phase. The total concentration of milk proteins (caseinates and serum proteins) in the water phase was 0.01%.

Interfacial Tension Measurements

The interfacial tension of oil/water systems containing varying concentrations of GMS in the oil phase is shown in fig. 1 as a function of decreasing temperature from 40°C to 1°C. At a concentration of 0.03% GMS or more, a decrease in interfacial tension (γ) takes place below a given temperature. The break point of such a curve can be referred to as the temperature at which the γ-decrease begins (T_γ). Fig. 1 shows that T_γ increases with an increasing concentration of GMS in the oil from 3°C at 0.03% to 28°C at 0.3% GMS.

Monoglycerides containing unsaturated fatty acids show less variation in γ when the temperature is decreased than GMS, as demonstrated in fig. 2.

The effect of the fatty acid chain length of monoglycerides on T_γ is shown in fig. 3. Medium-chain monoglycerides such as glycerolmyristate (GMM) show a T_γ break point at 8°C, whereas monoglycerides with longer chain lengths show T_γ break points at higher temperatures such as 19°C, 27°C and 39°C for GMP, GMS and GMB respectively. The effect of temperature on interfacial tension is thus more pronounced with long-chain monoglycerides than with short-chain monoglycerides. Furthermore, the relative decrease in interfacial tension (Δ_γ), when the temperature is lowered from 40°C to 5°C, was more pronounced in GMP, GMS and GMB than in GMM.

Anionic organic acid esters of monoglycerides, such as diacetyl tartaric acid esters (DATEM), are used in various food emulsions due to their hydrophilic properties. Their interfacial behavior as a function of temperature compared to that of GMS and sodium stearate (Na-C18) is demonstrated in fig. 4. DATEMs with

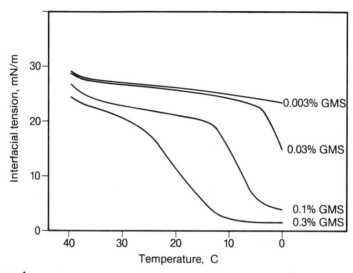

Figure 1
Effect of temperature on interfacial tension of oil/water
containing various concentrations of GMS added to the oil phase.

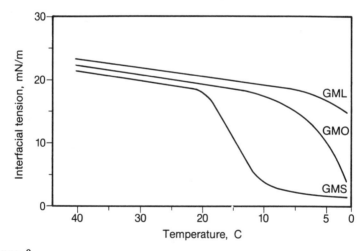

Figure 2
Interfacial tension versus temperature of oil/water containing
0.15% GMS, 0.15% GMO or 0.15% GML added to the oil phase.

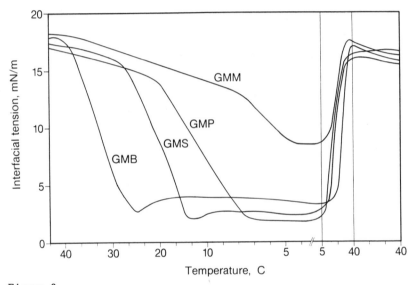

Figure 3
Interfacial tension versus temperature of oil/water containing
0.2% monoglycerides in the oil phase. GMM =
Glycerolmonomyristate, GMP = Glycerolmonopalmitate, GMS =
Glycerolmonostearate, and GMB = Glycerolmonobehenate.

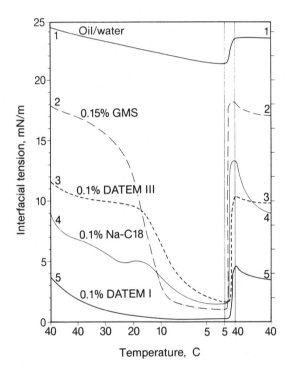

Figure 4
Interfacial tension versus temperature of oil/water interfaces
containing 0.15% GMS, 0.1% Na-C18 (sodium stearate), 0.1% DATEM
I, and 0.1% DATEM III in the oil phase.

one acyl group (DATEM I) are more hydrophilic than DATEMs with
two acyl groups (DATEM III). At 40°C there is a big difference in
the interfacial tension of oil/water systems containing 0.15% GMS
or 0.1% of the anionic compounds. DATEM I is the most effective
for lowering γ, followed by Na-C18 and DATEM III. However, when
the temperature is decreased to 5°C, all emulsifiers lower the
interfacial tension to approx. 2 mN/m, with DATEM I giving the
lowest value.

When 0.01% milk proteins are present in the water phase the
interfacial tension is 9.5 mN/m, and is not affected by
decreasing the temperature from 40°C to 5°C, as demonstrated in
fig. 5. After adding 0.2% GMP to the oil phase and 0.01% milk
protein in the water phase, the interfacial tension is not
changed at 40°C, but when decreasing the temperature slowly to
5°C a sharp decrease in interfacial tension takes place at about
18°C. At 10°C the value of γ is 2 mN/m, decreasing to a final
value of 1.5 mN/m at 5°C. During reheating from 5°C to 40°C the
interfacial tension increases to 9 mN/m, and reaches an
equilibrium of 8 mN/m at 40°C. The GMP/water curve (A) in figure
5 shows the interfacial tension of 0.2% GMP in the oil phase
compared to distilled water without proteins. The GMP/water curve
intercepts the γ-curve (B) with proteins present in the water
phase at the same temperature as the T_γ breakpoint (18°C) of
the GMP/protein curve (C). This demonstrates that monoglycerides
replace proteins at the interface at a temperature of 18°C or
below.

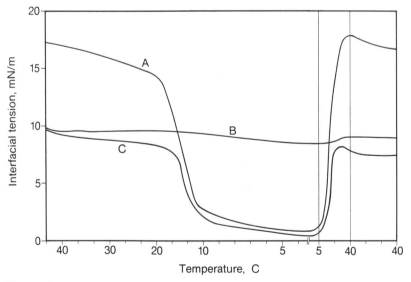

Figure 5
Interfacial tension versus temperature of oil/water interfaces.
A: 0.2% GMP in the oil phase against distilled water. B: Oil
without emulsifiers against a solution of 0.01% milk proteins in
water. C: Oil + 0.2% GMP against a solution of 0.01% proteins in
water.

Discussion and Conclusion

The formation of interfacial films (mono-layers) is strongly dependent on the solubility of surface active lipid monomers in the oil or water phase. In the case of low polar lipids, such as monoglycerides, the solubility in the oil phase is the determining factor for the formation of interfacial mono-layers. Since solubility is related to temperature, a relationship also exists between mono-layer formation and temperature.

The effect of temperature on mono-layers of propylene glycol esters, fatty acids, etc. was described by Lutton et al., in 1969 (3). They demonstrated a relationship between interfacial tension oil and water and the formation of solid condensed mono-layers. Our study of the behavior of monoglycerides is in agreement with the work of Lutton et al, and demonstrates that the surface activity of monoglycerides increases inversely with temperature. Furthermore, it has been shown that the transition from liquid condensed to solid condensed mono-layers of glycerolmonomyristate at air/water interfaces is determined by temperature (4). Such observations of the behavior of air/water mono-layers as a function of temperature correlate well with our studies of oil/water interfaces.

The adsorption of proteins to oil droplet surfaces in emulsions is based on the presence of hydrophobic segments of the protein molecules, which penetrate the outer triglyceride layers of oil droplets. The affinity of hydrophobic protein segments to lipids in the liquid state is much stronger than to crystallized lipid layers, which cannot dissolve protein. The formation of solid condensed mono-layers will therefore desorb proteins from the lipid surface of fat globules in emulsions at temperatures below the T_g point. Monoglycerides form solid condensed mono-layers at interfaces at low temperature, and this behavior may explain why monoglycerides enhance desorption of proteins from fat globules in e.g. ice cream mix during the aging process.

References:

1.   Goff, H.D.; Jordan, W.K. J. Dairy Sci., 1989, 72, 18.

2.   Barfod, N.M.; Krog, N.; Larsen, G.; Buchheim, W.
      Fat Science Technology, in press.

3.   Lutton, E.S.; Stauffer, C.E.; Martin, J.B.; Fehl, A.S.
      J. Coll. Interface Sci., 1969, 30, 283.

4.   Larsson, K.; Krog, N.; Riisom, T.
      In Encyclopedia of Emulsions; Becher, P. Ed.; Marcel
      Dekker: New York. Vol. II, p. 338.

RECEIVED September 9, 1990

Chapter 12

Function of α-Tending Emulsifiers and Proteins in Whippable Emulsions

J. M. M. Westerbeek[1], and A. Prins[2]

[1]Product Research and Analysis Laboratory, Campina-Melkunie,
P.O. Box 13, 5460 BA Veghel, The Netherlands
[2]Department of Food Science, Agricultural University, P.O. Box 8129,
6700 EV Wageningen, The Netherlands

In whippable emulsions a combination of proteins and small molecular emulsifiers is often applied as surface active agents in order to obtain desired whipping characteristics like e.g. firmness and volume. The consistency of these products is closely related to the aggregation of the dispersed fat globules. This study is performed on model systems, consisting of dispersions of the water insoluble α-tending emulsifier glycerol lacto palmitate (GLP) in sodium caseinate solutions. Though it is well known that the presence of relatively large amounts of so-called α-tending emulsifiers strongly promote aggregation of the fat globules the mechanism which explains this phenomenon is still unknown. The study included DSC, x-ray and neutron diffraction measurements to elucidate the micro-structure of the dispersed particles interfaces. Based on the obtained results a hypothetical model was developed explaining the colloidal stability of the fat particles in these whippable systems at different temperatures.

Whipped emulsions should always have solid-like properties, because the aerated product has to be stable against flow for a long period of time. In the case of whipped products a yield stress, caused by the presence of a particle network, is very often the way to prevent the product from flowing. The formation of such a network is frequently induced by the addition of small-molecular emulsifiers. Lipophilic, so-called α-tending emulsifiers, like propylene glycol monostearate (PGMS), acetylated monoglycerides (ACTM) or lactylated monoglycerides (GLP) are especially effective in promoting aggregation of fat globules (1,2). These α-tending emulsifiers may be characterized by the fact that they are non-polymorphic, can only exist in the α-crystalline form below the melting point of the hydrocarbon chains and are practically insoluble in water (2,3).

It is not understood what mechanism is responsible for this extensive fat particle aggregation phenomenon. Recently, Buchheim and Krog (4) suggested that crystallization of supercooled fat may play an important role in the occurrence of fat particle aggregation in whippable emulsions. It is indeed true that crystallization phenomena play an important role in the formation of a particle network in these emulsions. In whipped cream, the network probably consists of partly crystallized oil droplets. However, since whippable emulsions, which contain relatively large amounts of an α-tending emulsifier, do not churn during whipping (5), it seems likely that another mechanism is responsible for the instability of these emulsions.

It is well-known that mixtures of water and emulsifiers such as phospholipids (6,7) or monoglycerides (8,9) may form lamellar, cubic or hexagonal mesomorphic phases (10) above the crystallization temperature of the hydrocarbon chains of the emulsifier. When these systems are cooled down a lamellar gel phase may form. This gel phase is characterized by a lamellar structure of alternating layers of emulsifier and water molecules (10,11). The lipid molecules are crystallized in the α-polymorphic form. In Figure 1 three types of α-gel phase structures are presented (10).

Literature often state that α-tending emulsifiers do not show lyotropic mesomorphism (12-14). However the physical behaviour of GLP/water mixtures and glycerol monostearate/water mixtures is very similar below the crystallization temperature of the hydrocarbon chains of the lipid molecules (1,5). Depending on the amphiphile concentration, these systems both gel under these conditions, and show excellent whipping properties.

Our hypothesis is that α-tending emulsifiers are able to form an α-gel phase with water at the interface of fat particles below the crystallization temperature of the emulsifier hydrocarbon chains. The aggregated fat particles are linked to each other by this gel phase. In this paper a study on the physical behaviour of glycerol lacto palmitate (GLP), as an example of an α-tending emulsifier, is presented. Our aim has been to propose a satisfying mechanism for the structure formation in emulsions containing an α-tending emulsifier. The proper technique to study lipid polymorphism and mesomorphic behaviour is a combination of small and wide angle X-ray

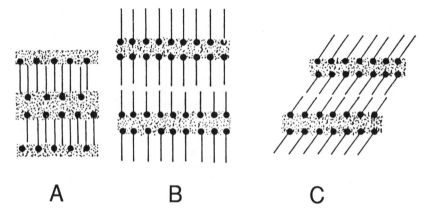

Figure 1 : The schematic structure of three possible lamellar gel phases formed by surfactants in contact with water below the crystallization temperature of their hydrocarbon chains.

diffraction (SAXD and WAXD) in addition to differential scanning calorimetry (DSC). Furthermore, neutron diffraction has proved to be a useful technique for structural analysis of mesomorphic phases as was the case for phospholipids (15-17).

EXPERIMENTAL

Materials.   Glycerol lacto palmitate (GLP) was purchased from Grindsted, Denmark. The sample is a very complex mixture of mono-, di- and triglycerides of lactic and palmitic acid. According to the supplier the product contains about 15% lactic acid. Sodium caseinate was obtained from DMV, Veghel, The Netherlands. This sample is a spray-dried milk protein in powder form containing 94.5% protein (N x 6.38) on moisture free basis, 5.2% moisture, 4.1% ash and 0.8% fat. Glucose syrup was obtained from Cerestar, The Netherlands. It is a combined acid/enzymatic hydrolysed corn starch which mean Dextrose Equivalent (DE) = 35.

Preparation of Spray-Dried GLP Powders.   Prior to spray-drying a concentrated emulsion of GLP particles in water was prepared containing 30% GLP (w/w), 5% sodium caseinate, 15% glucose syrup and 50% demineralized water. Hitherto the dry protein powder was dispersed directly in the melted emulsifier at 70°C. This protein/emulsifier mixture was added to the water phase at 70°C. Subsequently this dispersion was homogenized at a constant pressure of 100 atmosphere at 70°C in a high-pressure homogenizer (Rannie, 100 l/hr). This emulsion was spray-dried with an A/S NIRO Atomizer as described previously (18). Then the powders were stored at room temperature for DSC or diffraction studies.

Preparation of Freeze-Dried GLP Powders.   Melted GLP was mixed with demineralized water at a temperature of about 60°C with a Sorvall mixing apparatus. During this mixing procedure the sample was gradually cooled down to a temperature below the crystallization point of the emulsifier mixture. Then the obtained GLP gel was freeze-dried and stored at room temperature.

Differential Scanning Calorimetry.   DSC was performed with a Mettler TA-3000 system. In the DSC-cup the sample amount varied between 10 and 20 mg, depending on the concentration of potentially crystallizing matter present in the sample. The heating curves which are shown together in one Figure, were not corrected for differences in the amount of sample in the DSC-cup. Therefore the peak areas below the curves may not be interpreted as being representative of the heat values per unit amount of sample.

X-Ray Diffraction. SAXD-measurements have been performed with the spray-dried GLP powder samples. Part of
these samples were dispersed in demineralized water the
day before the x-ray diffraction experiments. The SAXD-
measurements were conducted with a Kratky camera, manu-
factured by A. Paar. The camera was equipped with a
Braun one-dimensional position sensitive detector which
was connected to a Braun multi-channel analyzer. The
radiation source was a PW-1729 x-ray generator, produ-
cing Ni-filtered CuKα-rays, its wavelength being 0.154
nm. Each channel of the multi-channel analyzer cor-
responded to a certain diffraction angle. The samples
were put into small glass capillaries (ϕ_{inside} = 1.0 mm)
with a wall thickness of about 0.01 mm. Calibration of
the apparatus was carried out with lead stearate.
Measurements were usually performed within 30 minutes.
Corrections were made for background noise and the curves were subsequently desmeared according to a program
based on Lake's theory (19).

Neutron Diffraction. Neutron diffraction experiments
were performed at ISIS, (Rutherford Appleton Laboratory,
Didcot, England). The small angle diffraction measurements were conducted with the LOQ spectrometer. The
spallation neutron source ISIS produces intense neutron
bursts 50 times per second by means of collisions of a
highly energetic, pulsed proton beam with a Uranium
target. The fast neutrons from the target station are
slowed down in a hydrogen moderator at a working temperature of 25 K in order to obtain a cold neutron enhanced thermal neutron spectrum. A rotating disk chopper at
25 Hz removes alternate pulses from ISIS to avoid frame
overlap from adjacent pulses. Neutrons of wavelengths
varying from 2 to 10 Å are recorded by a two-dimensional
position sensitive neutron detector. The area detector
is a multiwire (128x128), $^{10}BF_3$ filled, proportional
detector. It has 64x64 channels, its resolution being
1 cm in both directions, each one of them containing
about 80 time of flight channels. They are all handled
by an in-house data acquisition system and a Microvax
computer. Reduction of the raw LOQ's time of flight data
to a composite cross section I(Q) is done by accurate
transmission corrections over a wide range of wavelengths.
Dry and hydrated GLP powder samples, containing variable
amounts of D_2O were brought into circular shaped quartz
cuvets. The weight of the samples inside the cells was
determined in order to be able to calculate the average
cross section of each sample. The cell thickness could
be either one or two millimeters, depending on the a-
mount of D_2O and the amount of air present in each gel
sample. Each measurement lasted between 1 and 3 hours

for each sample depending upon whether $1*10^{+6}$ or $3*10^{+6}$ counts were required for acceptable data statistics.

RESULTS

Diffraction Studies.   The existence of α-gel phases can easily be characterized with diffraction techniques. The swelling of a crystal lattice caused by the uptake of water should clearly appear from an increase of the long spacing which can easily be measured with SAXD. The swelling of crystallized GLP was studied by means of hydration experiments with spray-dried GLP samples. To this end the powder, containing crystallized GLP particles with an average size smaller than 1 μm, was brought into contact with demineralized water at room temperature. After about 24 hours we performed SAXD experiments on both dry and wet samples. In Figure 2 two examples of SAXD spectra are represented. Comparing the two diffraction spectra it is obvious from the increase of the long spacing that hydration, though not complete, has indeed occurred at room temperature.

Comparable information may be obtained with neutron diffraction. In principal water molecules can be located within the structure of a multilayer if hydration is performed with D_2O (15,17). In order to obtain improved knowledge on the hydration properties of GLP, we performed neutron diffraction on three different types of GLP samples hydrated with variable amounts of D_2O. The first series was prepared from the spray-dried GLP-powder. The second series of experiments was performed with a freeze-dried GLP powder sample, free of any other additional component. The first and second series of samples were hydrated at ambient temperature. The third series of samples consisted of GLP gels, which were obtained by heating variable amounts of D_2O and GLP to a temperature of about 60°C in sealed bottles followed by cooling to ambient temperature under vigorous mixing. The differences between the three samples were thus related to composition, average particle size and temperature at which the samples were hydrated with D_2O.

The results of these measurements are represented in Figure 3. This Figure shows the measured long spacing of the gel phase of GLP as a function of the D_2O concentration in weight percentages. It is obvious that the values for the long spacings of the fully hydrated gel phases depend on the preparation method. The differences are not due to experimental error, because this Figure clearly shows that the values of the long spacing for sample B and C are quite constant at high D_2O concentrations. Our explanation for this effect is that fractionation of specific GLP components at the interface of the particles probably causes differences in the gel phase composition which, in turn, may cause

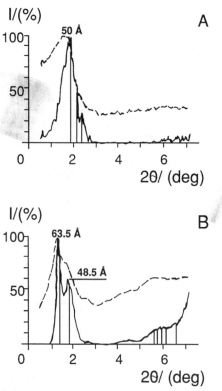

Figure 2 : SAXD curves obtained at 20°C for different
GLP samples.
A :   Dry, spray-dried GLP powder sample.
B :   Hydrated, GLP powder sample containing 67% water.
Both smeared (- - -) and desmeared (————) curves are
represented.

differences in the value of the long spacing at the point of maximum hydration.

Temperature Dependence. A proper technique to detect phase transitions in a system like mixtures of an α-tending emulsifier and water is differential scanning calorimetry. The influence of water on the phase behaviour of α-tending emulsifiers has been studied by means of both cooling and heating experiments (5). In Figure 4 the influence of the presence of water on the melting properties of spray-dried GLP is shown. It is obvious that the complete melting curve is shifted a few degrees towards a higher temperature, when the GLP particles are brought into contact with water at ambient temperature. This experiment confirms our hypothesis of hydration of α-tending emulsifiers below the crystallization temperature of the amphiphilic components hydrocarbon chains.

The heating curves suggest that a large part of the emulsifier molecules participates in the hydration process, because the shift observed almost accounts for the complete melting curve. This effect was not expected to occur so explicitly. If the gel phase would consist of only part of the total emulsifier sample, it would have been likely that the melting peak of the hydrated emulsifier, in comparison with the dry sample, would have been broader or would even have been split up in two peaks. Instead the endothermic heat peaks of the hydrated samples are significantly sharper than those of the dry samples which may be indicative of a better fitting of molecules with the bulky head groups into the crystal lattice of the α-gel phase.

Finally we will show that with temperature dependent SAXD measurements it can be proved that the long spacings we have shown indeed are related to a gel phase structure. In Figure 4 we showed that the main GLP fraction of the hydrated sample melts at a temperature of about 46°C, whereas the dry sample melts at a temperature of about 43°C. In Figure 5 the SAXD curves of the hydrated GLP powder at different temperatures are shown. These curves are not corrected in such a way that the peak heights can be compared with each other. The long spacing of about 62Å disappears at increasing temperature, indicating that the α-gel phase melts over a wide temperature range. At 47°C this long spacing is no longer present in the SAXD spectrum. Furthermore, this experiment gives evidence for the assertion that GLP does not show mesomorphism above the melting temperature of its hydrocarbon chains (12,14). The weak reflection at a d-value of about 48 Å is probably related to crystals formed by a relatively small part of the GLP mixture (also compare with Figure 4).

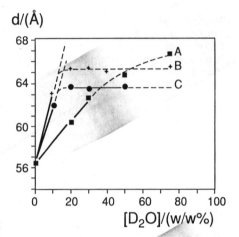

Figure 3 : The long spacing of three different GLP samples as function of the weight percentage deuterium oxide as determined with neutron diffraction at ambient temperature.
A :   Spray-dried GLP powder containing 60% GLP.
B :   Freeze-dried GLP powder containing 100% GLP.
C :   melted GLP.

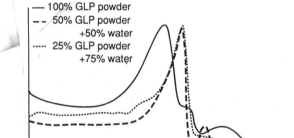

Figure 4 : Heating curves of completely crystallized dry and hydrated spray-dried emulsifier powders determined with DSC (Heating rate = 1°C/minute).

DISCUSSION

From the results shown in this paper it may be concluded without doubt that GLP is able to form a hydrated gel phase structure below the melting temperature of its hydrocarbon chains. The stability of this gel phase is excellent. We have proved with wide angle x-ray diffraction that the α-polymorphic form of both dry and hydrated GLP samples is very stable between 4 and 25°C (5). This may seem to be contradictory to results obtained for the stability of the α-gel phase of other emulsifiers like distilled monoglycerides. The stability of the α-gel phase of monoglycerides like glycerol monostearate is relatively poor (9). Dependent on the conditions (pH, temperature, salt concentration) a transition of this gel phase will occur into the β-polymorphic form which is accompanied with water exclusion from the crystal lattice. It has been reported the stability of the gel phase of monoglycerides may be prolonged, when small amounts of fatty acids (20) or lecithin (21) are added to the system. Presumably the formation of mixed crystals is favourable for the stability of the α-gel phase of an amphiphile.
Crystallization in the α-modification is a prerequisite for the formation of gel phases. Krog and Lauridsen (22) assert that emulsifiers which are non-polymorphic and stable in the α-modification like sodium stearoyl lactylate or tetraglycerol monostearate may form gels with water exhibiting long-term stability. Therefore, it is clear that α-tending emulsifiers being non-polymorphic are potential gel phase forming amphiphiles. The stability of the α-polymorphic form of GLP is probably determined mainly by steric repulsion of the bulky head groups of the mono- and diglycerides and the lactate esters of these molecules. Hydration of the polar head groups may provide an additional stabilizing effect which will be discussed later.
The structural parameters of the α-gel phase of GLP have not been discussed yet. It is not known which amphiphillic components participate as structure elements of this gel phase structure. Probably it contains both mono- and diglycerides and lactated esters of these components. We have shown that the long spacing of the hydrated gel phase is dependent on the preparation method. Therefore α-gel phase in fact is not a completely correct term for the hydrated crystalline structure of GLP. However for reasons of simplicity we will define this structure as being an α-gel phase. Figure 6 represents a hypothetical schematic molecular structure for the gel phase of GLP, which was brought in contact with water above the transition point of the emulsifier. In this model it is assumed that the dry bilipid layer thickness amounts to a value of approximately 55 Å. If the dry bilipid layer

INT/(−)

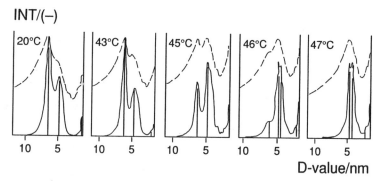

Figure 5 : X-ray diffraction curves of hydrated spray-dried GLP powders at increasing temperature. Both smeared (- - -) and desmeared (———) curves are represented.

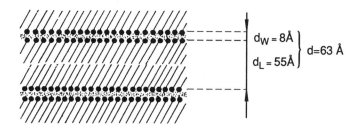

Figure 6 : A schematic molecular model for the α-gel phase of glycerol lacto palmitate hydrated above the crystallization temperature of the emulsifier.

thickness is correct, the water layer thickness of the gel phase would only be 8 Å or 9 Å. In the case of spray-dried GLP samples, which are hydrated at room temperature, the value for this water layer thickness probably is a few Angstroms larger.

DSC-measurements have indicated that the melting temperature of GLP is shifted towards a higher temperature in the presence of water. Such an effect has been observed with several other amphiphile/water mixtures (23,24) On the other hand, the melting temperature of monoglycerides (22) or phospholipids (25) decreases in the presence of water. It is obvious that the melting temperature of crystallized hydrocarbon chains of an amphiphile depends not only on the water concentration and hydrocarbon chain length but especially on the nature of the polar head group. In our opinion hydration of such a group may cause both an increase or a decrease in melting temperature of the chains, dependent on whether hydration leads to an increase or a decrease in the lattice energy of the crystalline amphiphile. In the case of GLP both a decrease in steric head group repulsion between molecules in opposite layers and formation of hydrogen bonds may play an important role in the melting point increase.

What does this phase behaviour of GLP mean for the stability of emulsions which contain both proteins like sodium caseinate and an α-tending emulsifier ? In the case of GLP containing emulsions we suggest a temperature dependent model for the structure of the o/w-interface of the emulsion droplets as represented in Figure 7. At a temperature exceeding the melting point of the amphiphile mixture the emulsion is stabilized by the proteins and perhaps by the relatively more hydrophilic components of the added emulsifier. Dependent on its concentration below the melting point of the emulsifier an α-gel phase is formed at the o/w-interface. It is likely that the stabilizing proteins will at least partly desorb spontaneously from the o/w-interface (26) and subsequently flocculation will then occur.

The particles will not coalesce because of the stabilizing hydration force exerted by the hydrated emulsifier molecules. The last ten years a better insight into hydration forces has been obtained (27). Hydration forces are only active over a very short range of about 30Å. If electrostatic repulsion is of no importance, thus in the case of neutral amphiphiles or ionics at high salt levels, amphiphiles may form mesomorphic structures as a result of van der Waals forces. It is believed that the van der Waals forces are counterbalanced by the short range hydration forces. In this respect it is significant to mention that it has been shown that hydration forces may form a strong barrier to both bilayer aggregation and fusion of phospholipid membranes

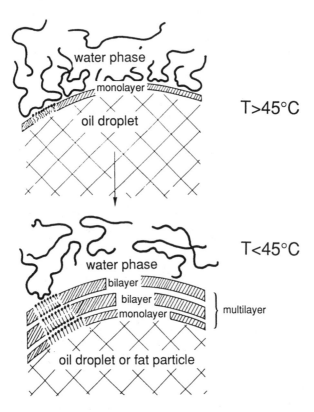

Figure 7 : Schematic model of the temperature dependent α-gel phase formation at the o/w-interface of fat
particles in an emulsion which contains both sodium
caseinate and GLP.

(28,29). Presumably the α-gel phase of GLP is also stabilized by this repulsive hydration force. Furthermore
it is possible the flocculated particles in a whipped
emulsion, containing a relatively high level of GLP on
fat basis, do not coalesce as a result of these
repulsive forces.

Acknowledgments.   The authors gratefully acknowledge
the help provided by P. Aarts (Unilever, Vlaardingen) in
carrying out the SAXD experiments, Dr J. van Tricht and
Dr. T. Rekveldt (IRI, Delft) for their advice and help
with the performance of neutron diffraction and J. van
Rijswijck (Campina-Melkunie, Veghel) in performing the
DSC-measurements.

LITERATURE CITED

1.   Krog, N. J. Am. Oil. Chem. Soc. 1977, 54, 124.
2.   Andreasen, J. Deutsche Molkereizeitung 1981, 36,
      1161.
3.   Krog, N. In Water Relations of Foods; Duckworth,
      R.B., Ed.; Academic: London, 1975.
4.   Buchheim, W.; Barfod, N.M.; Krog, N. Food Micro-
      structure 1985, 4, 221.
5.   Westerbeek, J.M.M. Contribution of the α-Gel Phase
      to the Stability of Whippable Emulsions; Ph.D.
      Thesis, University of Wageningen, 1989.
6.   Luzatti, V. In Biological Membranes; Chapman, D.,
      Ed.; Academic: London, New York, 1968, p. 71.
7.   Williams, R.M.; Chapman, D. In Progress in the Che-
      mistry of Fats and Other Lipids; Holman, R.T., Ed.;
      Pergamon Press: Oxford, 1970; Vol. 11, Chapter 1,
      p. 3.
8.   Lutton, E.S. J. Am. Oil. Chem. Soc. 1965, 42, 1068.
9.   Krog, N.; Larsson, K. In Chem. Phys. Lipids 1968,
      2, 129.
10.   Tiddy, G.J.T. Physics Reports 1980, 57, p. 1.
11.   Vincent, J.M.; Skoulios, A. Acta Cryst. 1966, 20,
      432.
12.   Andreasen, J. The Efficiency of Emulsifiers in
      Whipped Topping, Lecture at the International
      Symposium on Emulsions and Foams in Food Technolo-
      gy, Ebeltoft, Denmark, 1973.
13.   Boyd, J.V.; Krog, N.; Sherman, P. In Theory and
      Practice of Emulsion Technology; Academic: London,
      1974; p. 99.
14.   Schuster, G. Emulgatoren für Lebensmittel;
      Springerverlag: Berlin, Heidelberg, New York,
      Tokyo, 1985.
15.   Büldt, G.; Gally, H.U.; Seelig, J.; Zaccai, G. J.
      Mol. Biol. 1979, 134, 673.

16.   Worchester, D.L. In Biological Membranes; Chapman,
      D.; Wallach, D.F.H., Ed.; Academic: London, New
      York, San Francisco, 1976; Vol. 3, chapter 1, p. 1.
17.   Worchester, D.L.; Franks, N.P. J. Mol. Biol. 1976,
      100, 359.
18.   Barfod, N.M.; Krog, N. J. Am .Oil .Chem. Soc. 1987,
      64, 112.
19.   Lake, J.A. Acta Cryst. 1967, 23, 191.
20.   Krog, N.; Borup, A.P. J. Sci. Fd. Agric. 1973, 24,
      691.
21.   Kurt, N.H.; Broxholm, R.A. US Pat. 3,388,999 1968,
      Eastman Kodak Co.: Rochester.
22.   Krog, N.; Lauridsen, J.B. In Food Emulsi-
      ons; Friberg, S.; Marcel Dekker, Ed.; New York,
      1976, p. 67.
23.   Lawrence, A.S.C.; Al-Mamum, M.A.; McDonald, M.P.
      Trans Faraday Soc. 1967, 63, 2789.
24.   Vringer, T. de. Physicochemical Aspects of Lamellar
      Gel Structures in Nonionic O/W Creams, Ph.D.
      Thesis, University of Leiden, 1987.
25.   Hauser, H. In Reversed Micelles; Luisi, P.L.;
      Straub, B.E., Ed.; New York, 1984, p. 37.
26.   Krog, N.; Barfod, N.M.; Buchheim, W. In Food Emul-
      sions and Foams; Dickinson, E., Ed.; Royal Society
      of Chemistry, 1987, p. 144.
27.   Israelachvili, J.N. Chemica Scripta, 1985, 25, 7.
28.   Parsegian, V.A.; Fuller, N.; Rand, R.P. In Proc.
      Natl. Acad. Sci. USA 1979, 76, 2750.
29.   Lis, L.J.; McAlister, M.; Fuller, N.; Rand, R.P.;
      Parsegian, V.A. Biophysical J. 1982, 37, 657.

RECEIVED August 28, 1990

Chapter 13

Evaluation of Stabilizers for Synthetic Vesicles and Milk Fat Globules under Drying Stress

Kevin D. Whitburn[1] and C. Patrick Dunne[2]

[1]Department of Chemistry and Food Science, Framingham State College, Framingham, MA 01701
[2]Development and Engineering Center, U.S. Army Natick Research, Natick, MA 01760–5020

Dipalmitoylphosphatidylcholine (DPPC) vesicles in aqueous buffer, prepared by sonication and controlled fusion, were characterized by differential scanning calorimetry (DSC) analysis. Size variations in vesicle preparations during fusion were analyzed by spectroturbidity measurements and laser scattering analysis. Inasmuch as the measured turbidity of a vesicle preparation was shown to be directly proportional to the concentration of the phospholipid in the vesicled form, the relative quantity of vesicles remaining after applying a drying and rehydration procedure was estimated from turbidity measurements taken before and after drying. A parameter, η_R, the rehydration efficiency, was defined in terms of turbidity changes to express the resilience of vesicles to the drying-induced stress. Values of η_R were determined for vesicles dried and rehydrated in the presence of candidate preservatives, including carbohydrates and metal ions. Disaccharides stabilized dried phospholipid vesicles better than monosaccharides; polyols were destabilizing at all studied concentrations, and metal ions were mildly stabilizing only at low concentrations. Extracts of milk fat globules (MFG) from whole milk were analyzed for their resilience toward drying-induced rupture in the absence and presence of candidate preservatives by the spectroturbidimetric method. Of the studied additives, maltodextrin M-100 (a hydrolyzed corn starch) was found to be the most effective stabilizer. Analogous studies of dried diluted whole milk samples, in which the major contributor to sample turbidity was due to the component MFGs, indicated that the best single stabilizing additive toward drying-induced rupture of MFGs is maltodextrin M-100.

A fundamental problem with milk and milk-based foods and beverages is that they have limited shelf-lives outside a refrigerator. Although spray-dried milk products having extended storage stability

are commercially available, they usually derive from skim milk and
can have an unacceptable flavor or mouth feel to many consumers
after they are reconstituted with water. Skim milk is used predominantly to produce dried milk because the prior removal of the milk
fat minimizes the subsequent nonenzymatic oxidative rancidity in the
dried residue.

The fat content in milk exists as a distribution of tiny fat
globules or microspheres, bounded by membranes made up of complex
phospholipids and proteins, dispersed throughout the aqueous medium
(1). During the drying of milk, the removal of water from contact
with the surface membrane of the milk fat globules can lead to their
rupture, which can then lead to the dispersion of the fat component
throughout the residual dried protein, carbohydrate, and mineral
components (2). The blending of the fat with the other milk
components facilitates the potential for oxidative decomposition of
the dried milk, especially in the more reactive membrane phospholipids, and leads to the production of off-flavors and odors (3).

From published studies of synthetic and natural membrane
vesicles, it is known that additives can protect the membranes of
component microspheres against drying-induced rupture and fusion
(4-6); the additives include mostly sugars and some common salts.
Trehalose, a disaccharide, has shown particularly promising
potential for maintaining the structural integrity of both natural
and synthetic membrane vesicles under drying stress (6-8). This
protection has been correlated with the ability of the hydroxyl
groups on sugar molecules to replace the stabilizing influence of
water molecules on the surface of the dried membrane microspheres
(4,5,8,9).

Changes in the surface structure of component vesicles in an
emulsion induced by drying or freezing can be accompanied by
clustering and fusion phenomena, and by the leakage of a
pre-encapsulated solute (4,10). Relatively sophisticated
experimental techniques have been used to monitor the effects of
such stresses on vesicle preparations, including fluorimetry (7,11),
gamma radiation counting (6), electron microscopy (6,7), differential scanning calorimetry (DSC) (5,7,8), dynamic light scattering
(5), and gel filtration (5,6,7).

In order to provide a basis for developing a better dried whole
milk or nonskimmed filled milk product, we have adapted a spectroturbidimetric technique (12) to assess the effects of additives on
maintaining the stability of freeze-dried aqueous preparations of
synthetic vesicles and milk fat globules (MFGs). In addition to the
potential benefit for improving the stability of dried whole milk,
this study may provide insights into protecting liposomes used for
active-agent delivery systems in pharmaceutical applications.

Materials and Methods

Additives. All additives used to increase the rehydration efficiencies of dried emulsions were reagent grade from a variety of commercial resources. PEG is polyethylene glycol. Maltodextrin M-100 is
a commercial hydrolyzed starch product supplied by Grain Processing
Corp., Muscatine, IA; M-100 consists of 89% tetrasaccharides and
higher, 6% trisaccharides, 4% disaccharides, and 1% dextrose.

Preparation of FUV. This preparation is based on those of Barenholz
et al. (13) and Wong et al. (14). About 125 mg of dipalmitoylphosphatidylcholine (DPPC; Sigma, 99% purity) was suspended in 2.5 mL of
25 mM HEPES buffer previously adjusted to pH 7.0 with
tetramethylammonium hydroxide in a 15 mL Pyrex centrifuge tube. The
suspension was warmed to 50 °C, vigorously vortexed several times,
and then sonicated at 50 °C with a Branson Sonifier 350. The
suspension was sonicated for 24 @ 2 min blasts with 1 min breaks;
translucent opalescence in the sonicated emulsion was obtained
between blasts #15-20.

The sonicated sample was then centrifuged at 15,000 rpm for
50 min at 5-10 °C in a Sorval RC2-B (23,000 x g). The top 3/4 of
the supernatant was carefully removed with a Pasteur pipet. At this
stage, the emulsion is in the form of small unilamellar vesicles
(SUV); this form could be stored at 45 °C for several days. For
further processing and curing, the SUVs were stored at 20 °C for two
days, and then at 4 °C for 3 days. This storage procedure facilitated the fusion of the SUVs into fused unilamellar vesicles (FUV).
The sample was then centrifuged at 15,000 rpm for 45 min at 5-10
°C, from which the top 3/4 of the supernatant was carefully removed
and then stored at 20 °C.

A stock preparation of FUVs was essentially stable to further
fusion for up to 3 weeks. Greater uniformity of the FUV samples
taken from an aging preparation was obtained by taking aliquots from
the top of the supernatant after the stock was centrifuged at 6,000
rpm for 20 min. Absorbance scans of the diluted stocks were highly
reproducible for batch variations. Samples of the FUV were analyzed
for phosphorus by the method of Morrison (15). Yields of the
phospholipid in the FUV form of 60-70% were obtained by this preparative method.

Preparation of Milk Fat Globules. This preparation is based on the
method of Horisberger et al. (16). Two 5 mL portions of fresh
unprocessed whole bovine milk purchased from a local dairy, were
centrifuged at 5,000 rpm (8,000 x g) for 5 min. The infranatant
skim portion of the milk was then removed and discarded. The
remaining cream layer was resuspended without vortexing in 5 mL of a
50 mM TRIS (Sigma) buffer at pH 7.4 containing 0.15M NaCl (Fisher).
The resuspension was then centrifuged at 5,000 rpm for 5 min. This
procedure of removing the infranatant and resuspending the cream in
TRIS buffer was then repeated twice. The final resuspended cream
sample was chilled for two hours in an ice bath. After centrifugation at 5,000 rpm for 5 min, the infranatant was carefully removed
from the tilted centrifuge tube with a Pasteur pipet, and then
centrifuged again at 5,000 rpm for 5 min. This infranatant was
carefully decanted into a clean tube, and then filtered through a
Whatman No.1 filter paper. This preparative procedure removes the
largest portion of the original MFG population. Absorbance scans of
the resulting filtrate gave values of A_{350} : A_{650} typically
2.1 - 2.3. Unlike the DPPC vesicles, the MFGs were quite stable to
chilling-induced fusion in an ice bath for studied durations of up
to two weeks; preparations of MFG emulsions were stored in a refrig-
erated ice bath.

Spectroturbidimetric Analyses. All the emulsions used in this study
were scanned for absorbance, actually turbidity, over the UV-visible
range in 10 mm semimicro quartz cuvettes with a Gilford Response™

Spectrophotometer, having the capability of scanning a reference and
up to 5 samples in rapid sequence. In all cases, the turbidity
maximum was kept to <1.0 to minimize multiple scattering effects.
    For the quantitative analyses of DPPC FUV samples, small correc-
tions were made to the measured turbidity values for cuvette
mismatch, and for any marginal absorbance due to an additive reagent
in the studied wavelength range. All buffers were filtered through
BioRad 0.45 um Prep Discs, and all drying bottles were rinsed free
of any detergent residue. New polypropylene microcentrifuge tubes
were washed thoroughly in hot water before being used. The cuvettes
were kept thoroughly cleaned.

<u>Laser Particle Size Analyses.</u> DPPC microemulsions were analyzed for
the size distribution of their component vesicles on a Nicomp Model
200 Laser Particle Sizer (Nicomp Instrument, Inc., Santa
Barbara, CA) operated in the distribution mode for vesicular
microspheres at 632.8 nm and 5 mW. The output of this instrument
provides the mean diameter of the vesicle size distribution, the
standard deviation, and the closeness of the fit of the experimental
distribution to the Gaussian ideal.

<u>DSC Analyses.</u> Differential scanning calorimetry of the aqueous
emulsions was performed on a Microcal MC-2 solution calorimeter
(Microcal Inc., Northhampton, MA) in the range 20-90 °C at a scan
rate of 90 °C/h. The calorimeter is operated by dedicated data
acquisition and analysis software that includes an option to
estimate Δ H change for a phase transition.

<u>Freeze-Drying.</u> Samples of microemulsions were freeze-dried in an
FTS Dura Stop™ Dryer, operated at a shelf-temperature of 20 °C at a
pressure of 100-300 mT. Sample volumes of 0.5-1.0 mL were dried in
45x25 mm glass bottles for 80-120 min. Once evacuation begins, the
samples freeze within 2-3 mins after supercooling to ca. -5 °C at a
pressure < 1T. The frozen samples cool to -25 °C after 6-7 min at a
pressure of ca. 100 mT, and then gradually warm to ca. 0 °C after
20 min; the major portion of the drying occurs at ca. 0 °C.

<u>Results</u>

<u>Morphology and Stability of DPPC Vesicles.</u> Preparations of DPPC
vesicles in the concentration range 0.1-0.2% in aqueous HEPES buffer
at pH 7.0 were analyzed by DSC in the 20-90 °C range. The DSC
profiles differ significantly according to the morphology of the
vesicle preparations, as shown in Table I. T_c values are the
maximum transition temperatures for the gel $\Rightarrow$ liquid crystalline
phase change, and Δ H values are the associated enthalpy changes,
expressed in the table as the ratio of Δ H at the higher transition

Table I. Solution DSC Characterization of DPPC Vesicles

Treatment	Structural Form	T_c(°C)	$\dfrac{\Delta H_{hi}}{\Delta H_{lo}}$
Vortexed emulsion	MLV	35,41	6:1
Sonicated microemulsion	SUV	38,41	1:1
Post-sonicated fusion	FUV	41	

temperature to that at the lower temperature. Repeated scan cycles
gradually changed the DSC profile of the SUVs into that more characteristic of FUVs. The MLVs were marginally affected by repeated
scanning. The profile for the FUVs was entirely unchanged by three
heat-cool cycles over the studied temperature range.

Quantitative Spectroturbidimetric Analyses of Vesicle Fusion.
Figure 1 shows the turbidity profile in the wavelength range
250-350 nm of a 1:40 dilution of a freshly prepared 3% DPPC SUV
microemulsion; the [DPPC] in the SUV preparations was determined by
the Morrison analysis for phosphorus (15). The smooth, featureless
profile of decreasing turbidity with increasing wavelength is
characteristic of light scattering by the SUV.

   In order to accelerate vesicle fusion, freshly prepared SUV
microemulsions were placed in an ice bath. Over a duration of at
least 8 h, aliquots of the SUV were periodically diluted with HEPES
buffer and scanned in the 250-350 nm range on the spectrophotometer.
Figure 2 shows a profile of the turbidity at 250 nm normalized for
the mM concentration of DPPC against the ratio of the turbidities at
250 and 350 nm, τ_{250}/τ_{350} , for three separate SUV preparations. As
vesicle fusion proceeds, turbidities increase nonuniformly across
the wavelength range, with relatively greater increases at higher
wavelengths. Consequently, vesicle fusion is characterized by a
concomitant growth of turbidity at 250 nm and a decrease in τ_{250}/τ_{350}.

   In a parallel study, 1:10 dilutions of a fresh SUV preparation,
60 mM in DPPC, were analyzed by a Laser Particle Sizer during fusion
at 0 °C. In addition to the expected increase in the mean diameter
of the vesicles, the size analysis also indicated that the standard
deviation increases uniformly, while the degree of Gaussian fit for
the size distribution remains unchanged. The mean diameters of the
fusing SUV are profiled in Figure 3 against the τ_{250} / τ_{350} ratio
obtained from the simple spectrophotometric analysis.

   The turbidity of an emulsion also depends on the concentration
of particles in the sample. In Figure 4, the measured turbidities
of FUV samples at 250 and 350 nm are plotted against [DPPC].
Turbidity is directly proportional to [DPPC] up to the studied limit
of 5.2 mM at both wavelengths, indicating that multiple scattering
does not occur in this concentration range.

Rehydration Efficiency of Dried DPPC Vesicles. Because of their
inherent stability, FUVs were chosen for the drying studies. FUV
samples containing ca. 50 mM DPPC were centrifuged at 6,000 rpm
(9,000 x g) for 20 min at 15-20 °C. The top of the supernatant was
removed and used to prepare samples containing ca. 2 mM DPPC FUV in
HEPES buffer at pH 7, which were UV-scanned. After scanning,
1.00 mL portions were freeze-dried for 80 min in glass bottles. The
dried samples were rehydrated with 0.98 mL of distilled water by
gentle swirling for ca. 5 min; suspended particles of (presumably)
unvesicled DPPC, produced by the drying-induced rupture and coagulation of the original FUVs, were visible.

   The rehydrated samples were then centrifuged at 10,000 rpm
(15,000 x g) for 30 min at 15-20 °C to spin down the large
unvesicled particles; 0.65 mL of the supernatant was then carefully
removed from the microcentrifuge tube. The supernatant was
UV-scanned on the spectrophotometer. The effect of the centrifugation alone on the UV-profile of undried samples of the FUV was also
measured.

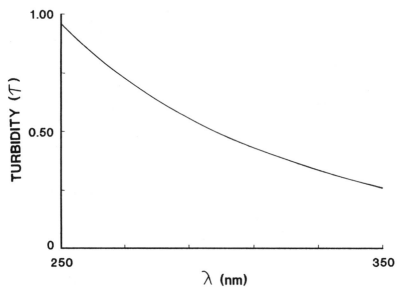

Figure 1.   Turbidity-wavelength profile of 0.075% DPPC SUV
microemulsion.

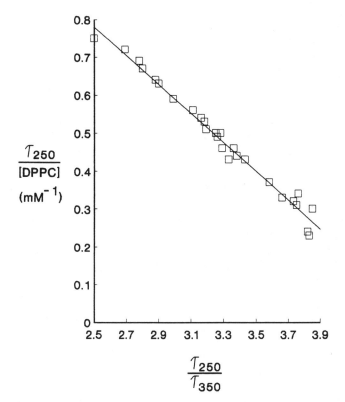

Figure 2. Turbidimetric characteristics of fusing DPPC vesicles; DPPC concentration normalized to 1.0 mM. τ_{250} and τ_{350} are the measured turbidities at 250 and 350 nm, respectively.

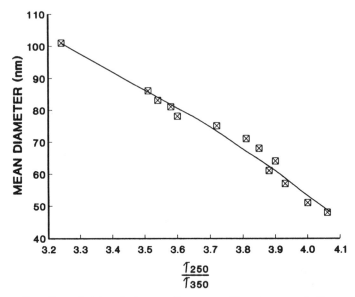

Figure 3. Correlation between the mean diameters and the τ_{250}/τ_{350} ratio of fusing DPPC vesicles.

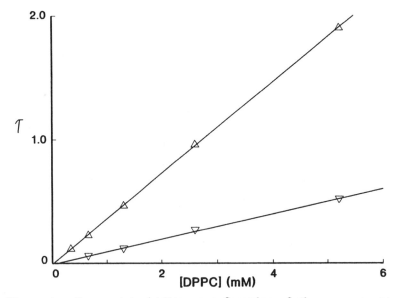

Figure 4. Measured turbidity as a function of the concentration of DPPC FUVs at 250 (△) and 350 (▽) nm.

As a result of the drying and rehydration procedure, the turbidity profile in the 250-350 nm range changed relative to that obtained before drying. Both τ_{250} and τ_{250}/τ_{350} were diminished significantly by the processing. As indicated in Figure 2, vesicle fusion alone, characterized by a reduction in the turbidity ratio can change τ_{250}. By measuring the turbidity ratio of the undried and dried DPPC vesicles, after centrifugation, reference to Figure 2 can provide the factor by which τ_{250} increases due to drying-induced fusion alone. In addition, the extent of vesicle fusion can be quantified from Figure 3; for DPPC FUVs having τ_{250}/τ_{350} ca. 3.7 initially, that of the reconstituted dried analogs decreases to ca. 3.3, corresponding to an increase in the mean diameter of the vesicle distribution from ca. 75 nm to ca. 95 nm.

Once τ_{250} measured after rehydration is corrected by the fusion factor, its value correlates with [DPPC] (Figure 4) in a slightly larger reconstituted form of FUVs. The ratio of τ_{250} of a centrifuged rehydrated sample to that of a centrifuged undried sample provides a measure of the rehydration efficiency, η_R, of the dried DPPC FUV. That is,

$$\eta_R = \frac{[\text{DPPC FUV(dried)}]}{[\text{DPPC FUV(undried)}]} - \frac{\tau_{250}(\text{dried})}{\tau_{250}(\text{undried})} \tag{1}$$

where τ_{250}(dried) is corrected for the increase due to drying-induced fusion, and τ_{250}(undried) is the value for the corresponding centrifuged undried sample. The value of η_R provides the extent of maintaining the original form of the vesicle population during the stress of drying and rehydration. A value of $\eta_R = 0.18(\pm0.03)$ was obtained for the dried and rehydrated DPPC FUV, independent of batch-to-batch variations of the stock used. With 10 mM sucrose present in the vesicle sample, the corresponding value of η_R increased significantly to 0.32 (±0.04).

In parallel studies on dried and undried samples of FUV in the absence and presence of 10 mM sucrose, the [DPPC] values in expression (1) were determined both by the Morrison analysis for phosphorus (15) and by DSC analysis, before and after the processing procedure. The Morrison analysis provides direct input into expression (1), while the DSC approach depends on the direct relationship between the value of ΔH for the FUV phase transition at 41 °C and the amount of DPPC in the sample. Table II shows the comparison of the values of η_R determined by the turbidimetric, phosphorus-content, and DSC methods. The acceptable correlation between these methods justified further application of the spectroturbidimetric method.

Table II. Comparison of Rehydration Efficiencies by
Three Methods

Method	η_R	
	no sucrose	10 mM sucrose
Turbidimetry	0.18 (±0.03)	0.32 (±0.04)
P-analysis	0.16 (±0.03)	0.38 (±0.05)
DSC	0.19 (±0.02)	0.30 (±0.02)

Effects of Additives on Resuspension of Dried DPPC Vesicles.
Samples of 2-3 mM DPPC FUV in HEPES buffer at pH 7.0 were dried and
rehydrated as described above in the presence of additives offering
potential protection against drying-induced vesicle rupture. The
values of η_R were determined spectroturbidimetrically using expression (1) for the broad range of additives listed in Tables III and
IV. A profile of η_R as a function of [sucrose] is shown in
Figure 5; η_R increases uniformly up to a plateau value of ca. 0.6
for [sucrose] > 40 mM, i.e. up to 12 mole sucrose/mole DPPC.

Profiles of η_R against [additive] were also obtained for the
additives in Table III, which shows the plateau (or peak) values
of η_R and the [additive] at which η_R is > 95% of its plateau value.
The effectiveness of an additive in protecting the FUV against
drying-induced fusion can be assessed by the tendency for η_R to
approach its limiting value of 1.0 at a minimal value of [additive].

In addition, several other additives were compared for their
ability to increase η_R of dried DPPC FUV at a single concentration
value, rather than over a range of concentrations. Table IV shows
the values determined at 15 mM of each additive.

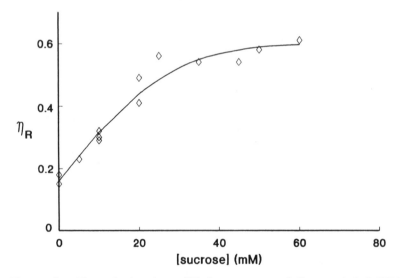

Figure 5. The rehydration efficiency, η_R , of freeze-dried DPPC
FUVs (2-3 mM) in the presence of increasing sucrose concentration.

Table III. Plateau Values of η_R and [Additive] at which η_R is >95% of the Plateau Value for Dried DPPC FUV. ([DPPC FUV] is 2-3 mM).

Additive	η_R plateau	[additive] (mM)
none	0.2	-
sucrose	0.6	40
trehalose	0.8	60
lactose	0.7	40
maltotriose	0.8	60
raffinose	0.7	>65
glucose	0.6	>65
xylose	0.6	>65
xylitol	0.2	a
glycerol	<0.1	b
maltodextrin M-100	0.5	5, c
Ca^{+2}	0.5	10, c
Mg^{+2}	0.6	10, c

a. η_R is independent of [xylitol] up to 60mM
b. no plateau is reached; glycerol is strongly fusogenic
c. further increases in [additive] reduce η_R from its peak value.

Table IV. Values of η_R for Freeze-Dried 2-3 mM DPPC FUV in the Presence of 15 mM Additive

Additive	η_R
none	0.18 (±0.04)
sucrose	0.38*
trehalose	0.43*
lactose	0.45*
maltose	0.40 (±0.04)
maltotriose	0.44*
raffinose	0.41*
glucose	0.28*
xylose	0.28*
fructose	0.38 (±0.05)
arabinose	0.36 (±0.06)
PEG 200, 400, 1000	<0.1
mannitol	0.25 (±0.03)
sorbitol	0.26 (±0.01)
Zn^{2+}	0.51 (±0.05)
betaine	0.25 (±0.05)
hydroxyproline	0.40 (±0.03)
alanine	0.23 (±0.06)
ß-alanine	0.17 (±0.03)
serine	0.23 (±0.01)
hydroxylysine	0.26 (±0.01)
sarcosine	0.31 (±0.06)

* from plots analogous to Figure 5; see Table III.

Effects of Additives on Resuspension of Dried MFG.   Emulsions of MFG
in TRIS buffer at pH 7.4 were prepared from fresh, unprocessed whole
bovine milk.   In Figure 6, the turbidity profiles of an MFG sample
over the 350-650 nm wavelength range show characteristic light
scattering by the fat globules, analogous to the scattering profile
in Figure 1 for DPPC FUV.   The MFG are considerably larger than the
synthetic vesicles (1), however, and show a different scattering
profile.   Photometric measurements were not taken below 350 nm
because protein components absorb in this region.   The correlation
between τ_{350} and %MFG w/v in a sample was determined by a plot
analogous to Figure 4 for DPPC FUV.   A direct proportionality
between τ_{350} and %MFG up to $\tau_{350} \sim 0.8$ was obtained, beyond which
progressive deviation from linearity due to multiple scattering was
observed; all subsequent quantitative studies of the MFG emulsions
were conducted with $\tau_{350} < 0.8$.

Figure 6 shows the turbidity profile of a 0.4 mg/mL emulsion of
MFG taken both before and after the sample was freeze-dried and
rehydrated.   Unlike the rehydration procedure for DPPC FUV, that of
dried MFG involved only the readdition of the original volume of
water, followed by gentle swirling before the spectrophotometric
scan was taken.   Because no fat particles were visible in the
resuspension, no centrifugation step was needed.   In addition to a
33% decrease in τ_{350}, there is a marginal 2% decrease in τ_{350}/τ_{650}
induced by the drying procedure.   When the MFG emulsion is dried in
the presence of 25 mM sucrose, τ_{350} decreases only 18%, and the
turbidity ratio increases 4%.   Thus, not only does drying and
rehydration induce measurable quantitative changes in the spectroturbidimetric profile of the MFG, but also the presence of a
carbohydrate additive affects the induced changes.   These

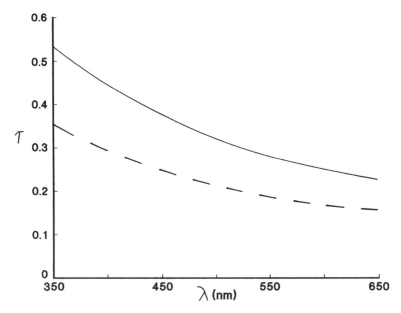

Figure 6.   Turbidity-wavelength profiles of a 0.4 mg/mL emulsion
of MFGs taken before (———) and after (----) freeze-drying and
rehydration.

qualitative features of dried rehydrated MFG provide a basis for assessing the effectiveness of protective additives in an approach analogous to that used for the DPPC FUV, in terms of η_R values.

Figure 7 shows a profile of $\eta_R = \tau_{350}(\text{dried})/\tau_{350}(\text{undried})$ as a function of added [sucrose], analogous to Figure 5 for DPPC FUV. Such profiles were also obtained for the additives listed in Table V, where the plateau values of η_R and the [additive] at which η_R is > 95% of its plateau value are presented.

Table V. Plateau Values of η_R and [Additive] at which η_R is >95% of the Plateau Value for Dried MFG. (%MFG ~0.04%)

Additive	η_R plateau	[additive] (mM)
none	0.6	-
sucrose	0.9	40
trehalose	0.9	40
lactose	0.9	20
glucose	0.9	>100
maltodextrin M-100	0.9	15
Ca^{2+}	a	N/A

a. increased $[Ca^{2+}] \geq 5$ mM decreases η_R

Effects of Additives on Resuspension of Diluted Milk. Samples of commercially available homogenized, pasteurized whole bovine milk, fortified with vitamin D, were diluted 1:250 with deionized water, and scanned in the spectrophotometer over the 350-650 nm wavelength range. As shown in Figure 8, a turbidity profile was obtained similar to that observed for MFG emulsions in Figure 6. When aliquots of the diluted milk were freeze-dried, simply rehydrated, and rescanned over the same wavelength range, a dramatic 75% decrease in τ_{350} was observed, in addition to a marginal 4% decrease in τ_{350}/τ_{650}. For the analogous experiment in the presence of 6 mM lactose, τ_{350} decreased by only 55%, and the turbidity ratio increased by 15%. As for the studies of MFG emulsions, these results suggest that the spectroturbidimetric method can be used to assess the effectiveness of added stabilizers in dried whole milk samples, again in terms of η_R values.

A 1:20 diluted whole milk sample in water was **further** diluted over a range of 1:10 to 1:67 either with water or 0.5N NaOH solution and left for 30 min with occasional vortexing. Each series of diluted milk samples was then scanned over the 350-650 nm range on the spectrophotometer. Inasmuch as 0.01N NaOH solubilizes casein micelles (17), any difference between the turbidities of the two series can be attributed to scatter by casein micelles. As shown in Figure 9, which profiles τ_{350} against the dilution factor, τ_{350} in the presence of NaOH is ca. 80% of that in its absence. This result indicates that most of the observed scatter at 350 nm is independent of casein and is due to the MFG component of the milk. Furthermore, the values of τ_{350} are linear with decreasing dilution up to $\tau_{350} = 0.8$.

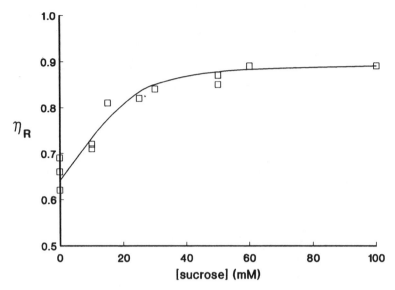

Figure 7. The rehydration efficiency, η_R , of freeze-dried MFGs (0.4 mg/mL) in the presence of increasing sucrose concentration.

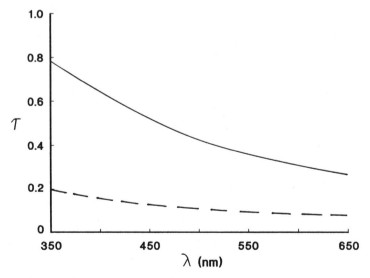

Figure 8. Turbidity-wavelength profiles of a 0.2% v/v diluted sample of whole milk taken before (——) and after (- - - -) freeze-drying and rehydration.

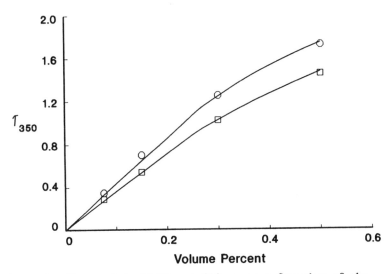

Υ_{350}

Figure 9.   Measured turbidity at 350 nm as a function of the volume percentage of milk in the absence (O) and presence (□) of NaOH solution.

A simple determination of η_R for dried, **undiluted** milk over a range of [additive] is automatically limited by the solubility of the additive at the levels needed, and by the inherent presence of a background [lactose] in milk of ca. 5% or 0.14M. Figure 10 shows plots of $\eta_R = \tau_{350}(\text{dried})/\ \tau_{350}(\text{undried})$ against log D, where D is the dilution factor or volume fraction of the milk samples, in the absence and presence of ca. 0.1M lactose. No lactose was added to the undiluted milk sample, and progressively more lactose was added to the more diluted milk samples to maintain the stated constancy of [lactose]. Measurement of τ_{350} values before drying required **further** dilution of the milk samples; less secondary dilution was required for the turbidity analysis of samples dried at higher primary dilution. Turbidity scans were determined on milk solutions diluted 400-fold overall before and after drying/rehydration. All dilutions were performed with deionized water, and allowance was made for the 12% solids content of the whole milk. Figure 11 clearly shows that η_R increases with increasing [lactose], and that the **same** effectiveness ($\eta_R \rightarrow 1.0$) is obtained at ca. 0.1M lactose, regardless of the extent of dilution of the dried milk sample.

The values of η_R against [lactose] were then determined at each of D=0.0032, 0.010, 0.032, and 0.10; due to the inherent lactose content of the milk, the minimum "default" value of [lactose] decreased with increasing dilution. From Figure 11, in which the profiles of η_R **vs** [lactose] for D=0.0032 - 0.10 overlay each other, it is not only evident that the same plateau value of η_R is reached at the highest [lactose], but also that the same η_R values are obtained at [lactose] values **before** the plateau is reached, regard-

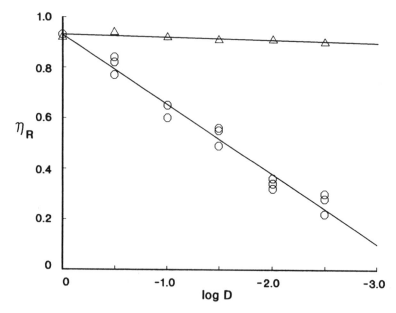

Figure 10. The rehydration efficiency, η_R , of freeze-dried
diluted milk samples against the log of the volume fraction,
log D, of milk in the sample, without and with added lactose; no
added lactose (O), 0.1 M lactose ($\triangle$).

less of the dilution factor. This overlay feature was observed
further for undiluted milk samples dried in the presence of
0.14 - 0.49M lactose; η_R ~ 0.97 was obtained at 0.49M lactose.

    That η_R of the dried milk samples depends on [lactose], and not
on the dilution factor (i.e., on the % MFG), suggests the same
degree of stabilization of the MFGs is offered by a given [lactose],
regardless of the concentration of the protected MFGs. That is, the
relative concentrations of lactose to MFG are unimportant over the
ranges of each component studied here. Consequently, the relative
protection of other stabilizers was studied at a nominal dilution
factor of 0.010, where the background [lactose] is a marginal
1.4 mM, and a large range of η_R from 0.3 - 1.0 is available for an
additive to show its effectiveness.

    Values of η_R as functions of [additive] were determined for
dried whole milk at a dilution factor of 0.010 for the additives
listed in Table VI. The profiles obtained for these additives were
characterized by increasing η_R to a plateau value with increasing
[additive], analogous to the profile in Figure 11. Table VI
contains the plateau (or peak) values of η_R , and [additive] at
which $\eta_R > 95\%$ of its plateau value for the dried diluted whole milk
samples.

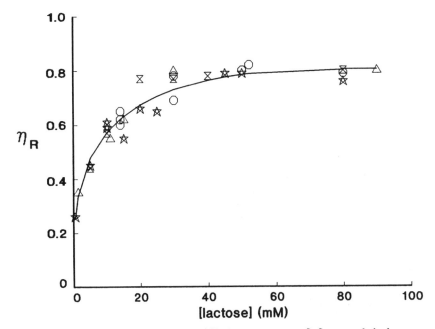

Figure 11. The rehydration efficiency, η_R , of freeze-dried
diluted milk in the presence of increasing lactose concentration
at different dilution factors, D; D = 0.0032 (★); D = 0.010 (△);
D = 0.032 (✕); D = 0.10 (○).

Table VI. Plateau Values of η_R and [Additive] at which
η_R is >95% of its Plateau Value for Dried MFG
in Diluted (0.010) Whole Milk.

Additive	η_R plateau	[additive] (mM)
none	0.3	-
maltodextrin M-100	0.9	15
lactose	0.9	35
trehalose	0.9	50
sucrose	0.9	25
glucose	0.9	30
glycerol	0.9	20
PEG 200	1.0	20
PEG 400	1.0	15, a
PEG 1000	0.8	20, a
sorbitol	1.0	20
Ca $^{2+}$	0.4	20, a

a. increased [additive] decreases η_R .

Discussion

DPPC Vesicles. Fused unilamellar vesicles of DPPC in aqueous buffer
can be prepared with a highly reproducible yield and size distribution by the technique described in this study. Although a preparation of FUV requires about a week to complete, the simplicity of the
technique allows at least two batches, chronologically displaced by
a few days, to be concurrently prepared without excessive consump-
tion of lab time. Although faster methods of preparing stable,
unilamellar phospholipid vesicles are available, their technical
complexity is generally greater (18,19). Once prepared, the FUV
stocks are storage-stable for several weeks at 20 °C; the storage
duration can be extended by diluting the stock preparation.
    Different precursor forms of the DPPC vesicles can be easily
extracted at earlier stages of the overall preparation and can be
characterized for their morphology and thermal stability by DSC
analysis, as shown in Table I. Although turbidimetric analyses of
the DPPC vesicles were performed in the wavelength range 250-350 nm
in this study, it is possible similarly to analyze vesicles prepared
from unsaturated synthetic phospholipids, or natural lecithins over
a wavelength range displaced towards the visible region where the
chromophore of the unsaturated lipid does not absorb. Chromophores
produced by peroxidation of the lipid may require an extension of
the analytical range to still longer wavelengths. The measured
values of τ_{250}/τ_{350} for the DPPC vesicles suggest that turbidity
varies inversely with wavelength to the power of 3.6 - 3.9, which
approaches the ideal $1/\lambda^4$ dependence associated with Rayleigh
scattering (20).
    The instability of DPPC vesicles in concentrated emulsions
towards chilling-induced fusion (14) has enabled the changes in the
size distribution of a vesicle population to be quantified by the
simple spectroturbidimetric method. From a single spectral scan of
diluted FUV sample, reference to Figures 2 and 3 enables an estimation to be made of both the [DPPC] in the microemulsion and the mean
diameter of the vesicle population, respectively. Although the
technique of spectroturbidimetry has been used previously to provide

relative sizings of fat globules in homogenized milk processing
(21), this technique has not been used to assess vesicle stabilizers
as it is applied in this study.

Resuspension of Dried DPPC Vesicles. The results in Table II
indicate that DPPC FUVs undergo substantial rupture and fusion as a
consequence of drying and rehydration stresses. In the absence of
an added stabilizer, only ca. 20% of the original FUVs are
resuspended in a size distribution comparable to that of the
pre-dried emulsion; the remaining DPPC has agglomerated into visible
particles, probably unvesicled, that are removed by centrifugation.
The extent of drying-induced rupture is diminished by the presence
of sucrose, a known stabilizer of dried vesicles (4). The quantitation of these drying effects is independent of the method used to
determine the relative amounts of DPPC in the emulsion before and
after the stress is applied. Since the turbidimetric technique is
substantially faster in providing this quantitation than the other
methods used, it provides a better means of determining the values
of η_R in expression (1). Furthermore, Figure 3 provides an estimation of the actual increase in the size of the resuspended FUVs,
which cannot by provided by the other methods. However, the
turbidimetric method does not indicate whether the residual FUVs
present after the processing have escaped rupture, or have reformed
after prior rupture.

The form of expression (1) indicates that $\eta_R \rightarrow 1.0$ as the
extent of drying-induced agglomeration of the vesicles decreases.
With the exception of glycerol and xylitol, all of the additives
listed in Table III increase η_R for the DPPC FUVs, indicating a
stabilizing influence against drying-induced damage. Of the
additives that protect the FUVs, trehalose and malto triose increase
η_R to the highest observed plateau value of ca. 0.8; however, as
much as 20 mole sugar/mole DPPC is required for this level of
protection. Lactose provides ca. 10% less protection for ca. 13
moles sugar/mole DPPC. Maltodextrin M-100 stabilizes strongly at
< 2 mole dextrin/mole DPPC, but destabilizes at higher relative
concentrations. Similar reversals of stabilizing influence were
observed for the metal ions Ca^{2+} and Mg^{2+}.

In the broader perspective, it appears that disaccharides offer
better protection than monosaccharides, and that simple polyols are
destabilizing. Metal ions protect only at low relative concentrations. This general trend has been previously reported (4). The
results in Table IV integrate with these trends for the most part;
the studied amino acids offer some protection against vesicle
rupture, but are not as effective in this regard as the disaccharides.

Resuspension of Dried MFG. MFG emulsions exhibit different spectroturbidimetric profiles from those of DPPC FUV samples. Although the
dependence of vesicle scattering on wavelength differs substantially
because the MFGs are at least 10 times larger than the FUVs (1),
both vesicle types exhibit a similar behavior in response to an
applied drying/rehydration stress; the turbidity of a rehydrated
sample is diminished significantly relative to an undried sample,
and this effect is progressively ameliorated by the addition of
known vesicle stabilizers at increasing concentration.

Of the additives listed in Table V, the maltodextrin is the most
effective stabilizer, followed by lactose; Ca^{2+} is moderately
destabilizing. Lactose is significantly more protective than the
other disaccharides, suggesting that some specificity of interaction
exists between the milk sugar and the milk fat globule membrane. As
for the DPPC FUVs, the disaccharides are more protective than the
monosaccharide.

Resuspension of Dried Whole Milk. The drying of dilute whole milk
leads to diminished τ_{350} in the observed spectroturbidimetric
profile of the rehydrated samples. The profile is again characteristic of scattering by the MFGs; scattering by casein micelles was
shown to be of marginal importance over the studied concentration
range (Figure 9). Figure 11 shows that the stabilization of the
MFGs in diluted whole milk by lactose depends only on [lactose] and
not on the dilution factor of the milk. That the degree of protection offered by this stabilizer is not dependent on the relative
concentration of these components, i.e., on the [lactose]:[MFG]
ratio, is difficult to interpret and requires further investigation.
    The results in Table VI indicate that maltodextrin M-100 is the
most protective of the listed additives towards drying-induced
modification of the MFG morphology in 1:100 diluted whole milk.
Glycerol, PEG 200, and sorbitol are marginally less effective; the
higher MW PEGs are destabilizing at concentrations > ca. 20 mM.
Lactose is more protective than trehalose, but less so than sucrose.
Ca^{2+} is destabilizing beyond 20 mM. These trends are broadly
similar to those obtained for dried MFG samples, in Table V, particularly for the best and worst stabilizers. Glycerol and sorbitol
exhibit opposite influences of stabilization and destabilization on
dried MFG and DPPC FUV, respectively. This contrast probably
reflects a favorable structural interaction of these additives with
the membrane proteins of the natural emulsion.
    It is notable that maltodextrin M-100 shows optimal stabiliza-
tion for the MFGs in Tables V and VI. Inasmuch as the comparisons
of effectiveness of the stabilizers are made on a per mole basis,
part of the advantage of the dextrin over the other carbohydrates
may be in its large MW; dextrins provide more OH-groups per mole
than the other carbohydrates. Not only does this dextrin offer
optimal protection, but also it has very low sweetness, high
solubility, and is not active in Maillard browning reactions with
protein components (22). All these features consolidate the choice
of maltodextrin M-100 as the best single stabilizer of dried whole
milk. As a secondary protectant, the background concentration of
lactose in milk offers some stabilization, but, as long commercial
experience with drying whole milk has demonstrated, the extent of
this self-protection is adequate only for short-term storage
stability of the dried product.

Acknowledgments

    The authors wish to thank the following summer research
assistants for their experimental contributions to this work:
Michael DiNunzio, Deborah Rodolfy, and Tad Sudnick. Ms. Bonnie
Atwood's help in preparing the figures is greatly appreciated, as is
Ms. Donna Walker's effort in typing the manuscript.

Literature Cited

1. Mulder, H.; Walstra, P., The Milk Fat Globule. Commonwealth Agricultural Bureau; England and Centre for Agricultural Publishing and Documentation; Netherlands, 1974.
2. Coulter, S.T.; Jenness, R.; and Geddes, W.F., Adv.Food Res. 1951, 3,45.
3. Brunner, J.R. Principles of Food Science; Fennema, O.R., Ed; Marcel Dekker; New York, 1976, p.619.
4. Crowe, J.H.; Crowe, L.M.; Carpenter, J.F.; Aurell Wistrom, C. Biochem. J. 1987. 242, 1.
5. Strauss, G.; Shurtenberger, P.; Hauser, H., Biochem. Biophys. Acta 1986, 858, 169.
6. Madden, T.D.; Bally, M.B.; Hope, M.J.; Cullis, P.R.; Schieren, H.P.; Janoff, A.S., Biochem. Biophys. Acta 1985, 817, 67.
7. Crowe, L.M.; Crowe, J.H.; Rudolf, A.; Womersley, C.; Appel, L.; Arch. Biochem. Biophys. 1985, 242, 240.
8. Crowe, J.H.; Crowe, L.M.; Chapman, D., Arch. Biochem. Biophys. 1984, 232, 400; 1985, 236, 289.
9. Anchordoguy, T.J.; Rudolf, A.S.; Carpenter, J.F.; Crowe, J.H., Cryobiology 1987, 24, 324.
10. Strauss, G., Membranes, Metabolism, and Dry Organisms; Leopold, A.C., Ed; Comstock Publishing Assoc: Ithaca, N.Y., 1986, p. 343.
11. Womersley, C.; Uster, P.S.; Rudolf, A.S.; Crowe, J.H., Cryobiology 1986, 23, 245.
12. Goulden, J.D.S. Brit. J. Appl. Phys. 1961, 12, 456.
13. Barenholz, Y.; Gibbes, D.; Litman, B.J.; Goll, J.; Thompson, T.E.; Carlson, F.D. Biochemistry 1977, 16, 2806.
14. Wong, M.; Anthony, F.H.; Tillack, T.W.; Thompson, T.E.; Biochemistry, 1982, 21, 4126.
15. Morrison, W.R., Anal. Biochem. 1964, 1, 208.
16. Horisberger, M.; Rosset, J.; Vonlanthen, M. Exp. Cell. Res. 1977, 109, 361.
17. Goulden, J.D.S., J. Dairy Res. 1960, 27, 67.
18. Szoka, F.C.; Papahadjopoulos, D., Annu. Rev. Biophys. Bioeng., 1980, 9, 467.
19. Lasic, D.D.; Belic, A.; Valentinic, T., J. Am. Chem. Soc., 1988, 110, 970.
20. Kerker, M., The Scattering of Light and Other Electromagnetic Radiation; Academic Press; New York, 1969.
21. Goulden, J.D.S.; Phipps, L.W., J. Dairy Res. 1964, 31, 195.
22. Grain Processing Corp.; technical literature.

RECEIVED August 16, 1990

Chapter 14

Interaction of Proteins with Sucrose Esters

Shio Makino and Ryuichi Moriyama

Department of Food Science and Technology, Faculty of Agriculture,
Nagoya University, Nagoya, Aichi 464–01, Japan

Food approved sucrose esters were interacted with a
variety of proteins differing in their molecular
surface to gather information about their
characteristics. Sucrose monoesters were found to
associate non-specifically with the hydrophobic
binding sites in native proteins. Comparing with other
nonionic surfactants, lauryl ester was found to
possess the power to extract membrane–associated
proteins from biological membranes. It efficiently
disturbs hydrophobic interactions between nonpolar
proteins, but not strong enough to induce changes in
the secondary structure of proteins.

The sucrose esters of fatty acids are approved food additives for
use as emulsifiers. In addition to the non-toxicity of the intact
esters and of their hydrolysis products, changes in the degree of
esterification with fatty acids of varying hydrocarbon chain lengths
can cover a wide range of hydrophilic–lipophilic balance values.
Antimicrobial activities of the esters further potentiate their use
in the food industry (1-3). Furthermore, recent studies show that
the esters possess the ability to efficiently solubilize hydrophobic
proteins from biological membranes (4-6). These esters thus
represent promising surfactants, emulsifiers and solubilizing agents
for aggregation in the fields of food biochemistry and food
processing.

There have been a number of reports concerning the effects of
sucrose esters on the properties of starch and wheat flour (7-10).
In spite of their increasing use, the interactions of the esters
with proteins in foodstuffs have drawn little attention. The
available information is limited to certain qualitative properties
of the esters such as their solubilization power. Sucrose itself
acts as a stabilizer of protein structure when introduced into a
solvent medium (11). This may suggest the possibility that esters
having sucrose as a hydrophilic moiety interact with proteins in a

manner differing from that of other nonionic surfactants. One objective of this study is to describe the possible modes of interaction of sucrose monoesters with proteins.

As noted above, it has been suggested that sucrose esters are superior to other nonionic surfactants for non-denaturing solubilization of membrane proteins (4-6). To obtain further information on the action of sucrose esters on hydrophobic proteins, we examined the effects of lauryl ester (SE12) on the state of association of a hydrophobic protein relative to the surfactant's solubilization power. We selected band 3 protein of bovine erythrocyte membrane, an anion transporter, for this purpose. This protein is present as a mixture of dimers and tetramers (12). It consists in fact of two distinct domains; the cytoplasmic 50kDa domain and the membrane-bound 58kDa domain. Oligomers of band 3 protein are formed by interactions within both domains (12,13). In order to evaluate only the action of SE12 on hydrophobically associated domain, we isolated the membrane-bound domain of bovine band 3 (58kDa fragment) and used it.

MATERIALS AND PROCEDURES

Sucrose esters of caprylic, capric, lauric, myristic and palmitic acids (referred to as SE8, SE10, SE12, SE14 and SE16, respectively) were supplied by Mitsubishi Kasei Co. The monoesters (purity > 98%) were used. C12E9 was obtained from Nikko Chemicals, Triton X-100 was from Wako Pure Chemicals. n-Dodecyl-N,N'-dimethylamine oxide was synthesized by the method of Lotan et al (14). Bovine serum albumin, fraction V, and ovalbumin, grade V, were obtained from Sigma. α_{s1}-Casein was prepared according to the method of Thompson and Kiddy (15). Membrane-bound chymotryptic fragment of bovine band 3 (58kDa fragment) was isolated by ion-exchange column chromatography (12).

Binding of SE8 and SE10 to bovine serum albumin and ovalbumin was measured by the method of equilibrium dialysis in two-chambered (1 ml each) Lucite cells separated by a Visking membrane. After equilibrium was attained (in 2 days below the cmc and 4-7 days near and above the cmc of the ligand), the concentrations of the esters were determined by the modified Anthron method (16). Maximum binding numbers to α_{s1}-casein and 58kDa fragment of band 3 were calculated from the excess amount of sucrose esters, which was co-eluted with the proteins from an anion-exchange column, as described previously (12).

Gel permeation chromatography of the SE8 and SE10 micelles was carried out at room temperature on a Sephacryl S-200 column which had previously calibrated using standard globular proteins with known molecular weights.

Circular dichroic spectra were recorded on a JASCO J-40 spectropolarimeter. A cell of 1 mm light path was used.

The molecular weight of protein moiety in a protein-surfactant complex was determined by sedimentation velocity technique, as described previously (12).

Extractability of the protein from bovine erythrocyte ghosts with nonionic surfactants was determined as follows. Washed and packed ghosts (0.5 ml) were mixed with suitable amounts of buffer and nonionic surfactant solution to bring the protein concentration to 4 mg/ml and to obtain the desired surfactant concentration. The suspensions were incubated for 20 min on ice and then centrifuged. The supernatant fractions were collected and their protein concentrations were analyzed.

RESULTS

<u>MOLECULAR CHARACTERISTICS OF SUCROSE MONOESTERS.</u> Data on the binding of surfactants to protein provide critical information required for understanding their mode of interaction. Experimentally obtainable quantities are the molal binding number and the concentration of free ligand in equilibrium with the complex; the resulting curve is called binding isotherm. Determination of these quantities allows calculating of the binding free energy and the maximum amount of binding. The binding isotherm can easily be constructed using a dialysis equilibrium method, provided that the surfactant used has a relatively high cmc. In order to choose sucrose monoesters appropriate to this purpose, we previously examined some of their relevant characteristics (17).

Results of size exclusion chromatography gave a Stokes radius (Rs) of 33 A for the micelle of SE12, a value close to that of bovine serum albumin (molecular weight 67,000). This indicates that the micelle of SE12 consists of roughly 120 surfactant molecules. On the other hand, the micellar molecular weight of SE10 was determined as 10,500 ± 4,000 by using both the observed Rs and the intrinsic viscosity of the SE10 micelle (15 ± 2 A and 2.1 ± 0.05 cm^3/g, respectively), assuming a globular micelle. Hence, SE10 forms micelles with an aggregation number of approximately 20. In fact, micellar solutions of SE8 and SE10 reached equilibrium within 3 days when dialyzed using dialysis membrane of molecular weight cut-off 12,000–14,000. The above observations suggest that, among monoesters evaluated here, SE10 is the most pertinent one for studying the interaction of proteins with a surfactant in both monomeric and micellar states, by the dialysis equilibrium method. Table I summarizes the molecular characteristics of SE10 and SE12 used in the present experiments. As noted, sucrose esters show a negative circular dichroic band centered around 215 nm, thus overlapping the bands induced by the polypeptide backbone of proteins.

Knowledge on the monomer concentration of surfactant above the cmc is of potential validity for the analysis of protein–surfactant interaction. In small micelles such as SE10, the concentration of monomeric surfactant might continue to increase above the cmc at which micelles are first detectable, until the bulk of the surfactant forms micelles. A numerical approximation was made by assuming that the micelle formation of SE10 is governed by the equilibrium 20(SE10) $\rightleftarrows$ (SE10)$_{20}$. The result of a such calculation, for a reasonable value of the equilibrium constant, 3.45×10^{46}, is shown in Figure 1. The equilibrium constant corresponds to a 2% fraction of the SE10 being in micellar form at the cmc. We have analyzed the binding of SE10 to bovine serum albumin by using this calculated monomer concentration.

<u>INTERACTION OF SUCROSE ESTERS WITH WATER–SOLUBLE PROTEINS.</u> In order to study the interaction of sucrose esters with water–soluble proteins, we selected ovalbumin, bovine serum albumin and α_{s1}-casein

Table I. Molecular Characteristics of Sucrose Monoesters at 25°C

	Critical micelle concentration (mM)	Ellipticity at 220nm (deg.cm^2/dmol)	Partial specific volume (cm^3/g)	Micellar molecular weight
SE10	2.52	−450	0.776	10,500
SE12	0.40	−430	0.798	62,000

(Reproduced with permission from Ref. 17. Copyright 1983 Maruzen.)

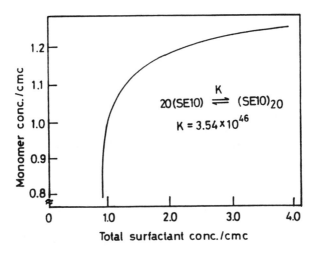

Figure 1. Relationship between monomeric concentration and critical micelle concentration of SE10 at 25°C. (Reproduced with permission from Ref. 17. Copyright 1983 Maruzen.)

as model proteins; these polypeptides differ in their molecular surface properties. Bovine serum albumin has been indicated to possess hydrophobic patches or crevices at the molecular surface in its native globular form (18). α_{s1}-Casein, which lacks a rigid ordered structure, harbors several primary structure areas consisting of hydrophobic amino acid alignments (19). Native ovalbumin does not have enough hydrophobic surface to accommodate hydrophobic ligands with high affinity (20).

Experimental data for the binding of SE8 and SE10 by ovalbumin and bovine serum albumin are shown in Figure 2 (17). This plot represents _ the number of ligand molecules bound per protein molecule ($\bar{\nu}$) as a function of the free ligand concentration (C) in equilibrium with the protein. Detail binding experiments were conducted for SE10. It was seen in the binding of SE10 to bovine serum albumin that the amount of bound surfactant increases monotonically with free ligand concentration, and that cooperative binding to very high values of $\bar{\nu}$, such as observed for alkyl sulfates and sulfonates (21), is not seen here. On the other hand, virtually no binding to sucrose esters is observed in the case of ovalbumin, a protein which lacks hydrophobic surfaces in its native state.

Data for the binding of SE10 to bovine serum albumin were analyzed by the procedure of Scatchard, as shown in the inset of Figure 2. The $\bar{\nu}$/C values above the cmc were calculated by using the monomer concentration shown in Figure 1, and the resulting points (filled circles in the inset) are in no way anomalous, suggesting that only the monomeric form of the surfactant can associate with bovine serum albumin. The Scatchard plot suggests that the binding can be accounted for, within experimental error, by 5 to 7 identical and non-interacting sites with an association constant of 500 to 900 liters/mol. The solid line in Figure 2 represents the theoretical binding isotherm as calculated using the following values: 6 binding sites and an association constant of 600 liters/mole. Similarly, the binding of SE8 to bovine serum albumin was reasonably accounted for by 5 identical, non-interacting sites with an association constant of 50 liters/mol.

Sucrose esters were found to bind to α_{s1}-casein as shown in Figure 3. The plot shows the effect of hydrocarbon chain length on binding at different temperatures. These measurements were taken using an ion-exchange procedure with surfactant solutions of concentration several fold higher than the cmc at 25°C. Thus, the amount of bound surfactant represents a maximum binding number. The protein was capable of binding the typical hydrophobic probe, BADS, as well as bovine serum albumin. The fluorescence of BADS, induced by its binding to the protein, followed a bell-shaped curve with a maximum intensity at 25°C for the temperature range from 4°C to 40°C. This behavior is closely parallel the temperature dependence of the binding of sucrose monoesters, clearly indicating that hydrophobic areas provide major binding sites for sucrose esters. Furthermore, as seen in Figure 3, the effect of chain length on binding suggests that the hydrophobic moieties of the binding sites are circumscribed and limited in the hydrocarbon chain length that can be accommodated. This effect is also observed for the binding of alkylsulfates on native bovine serum albumin (18). The small binding numbers observed at saturation binding levels again suggest that

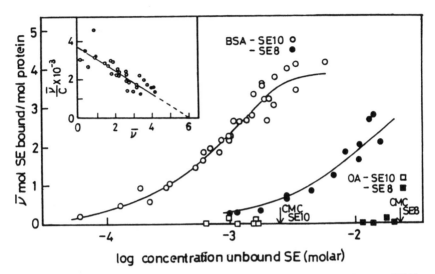

Figure 2. Binding of sucrose esters to bovine serum albumin (BSA) and ovalbumin (OV) in 0.1 M Na_2HPO_4–NaH_2PO_4 buffer, pH 7.2, at 24°C. (Reproduced with permission from Ref. 17. Copyright 1983 Maruzen.)

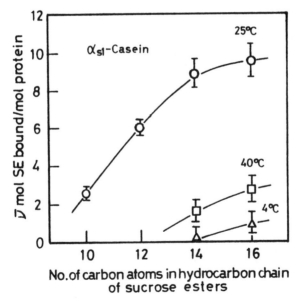

Figure 3. Binding of sucrose esters to α_{s1}-casein in 0.15 M NaCl, 10 mM Tris–HCl buffer, pH 8.0, at different temperatures.

sucrose   esters   fail   to   bind   cooperatively   to   proteins   with
accompanying conformational change.

EFFECT Of SE10 ON HEAT DENATURATION OF BOVINE SERUM ALBUMIN AND
OVALBUMIN. The binding of SE10 to bovine serum albumin did not
affect the secondary structure of the protein (17). This was
confirmed by the observation that the circular dichroic spectrum of
the protein–SE10 complex can be represented as the sum of the
individual circular dichroic intensities of the protein and of SE10
alone. The addition of SE10 to bovine serum albumin at a saturated
binding level increased its denaturation temperature from 50°C to
60°C. Further addition of the surfactant up to 60 mM did not promote
the stabilizing effect. A similar effect was observed for bovine
serum albumin complexed with other nonionic surfactants. However,
SE10 did not stabilize the native structure of ovalbumin.

EFFECT OF SE12 ON A STATE OF ASSOCIATION OF BOVINE BAND 3 PROTEIN.
Sucrose esters have recently received considerable attention as
solubilizing   agents   for   integral   membrane   proteins (1–4).
Solubilization power of SE12 to extract membrane proteins from
bovine erythrocytes was compared to those of several other types of
nonionic surfactants. The extractability by most of the surfactants
was nearly maximum above the cmc. Triton X–100 solubilized less than
10% of the total amount of protein from the membrane. On the
hand, SE12 and n–dodecyl–N,N–dimethylamine oxide extracted roughly
half of the total protein, or twice as much as that the value for
$C_{12}E_9$. Further, we have observed that SE12 selectively solubilizes
the membrane–bound 58kDa fragment of band 3 protein from
chymotrypsin–treated membrane, whereas $C_{12}E_9$ fails to do so.
    The potent solubilization power of SE12 may act to weaken in
fact hydrophobic interactions that support the oligomeric structure
of band 3 protein. This prompted us to characterize the molecular
state of the 58kDa fragment both in SE12 and $C_{12}E_9$ solutions.
Preliminary studies have shown that the attachment of DIDS, an anion
transport inhibitor, on the 58kDa fragment produces a single
molecular species (not a mixture of species) in its isolated state.
We therefore used the 58kDa fragment labeled with DIDS in the
following experiments.
    The state of association of this fragment was studied by gel
permeation chromatography, gel electrophoresis and sedimentation
velocity   measurements (22). Figure 4 shows   gel   permeation
chromatography patterns for DIDS–labeled 58kDa fragment in the
presence of 0.1% $C_{12}E_9$ or 0.1% SE12. Using a calibrated column, we
calculated that the complex with $C_{12}E_9$ has an Rs of 75 A, while SE12
formed a complex with an Rs of 58 A. A smaller SE12 complex was also
located during gel electrophoresis (22). It is our opinion that
these difference might relate to the particular molecular states
retained by the protein in SE12 and $C_{12}E_9$ solutions. This in turn
allowed us to determine the molecular weight of the polypeptide
chain moiety in the surfactant–protein complex, according to the
sedimentation velocity method applied to a multi–component system
(12). Table II summarizes the molecular weights of the protein
moieties obtained, and molecular parameters used in the
calculations.
    Results indicate that the 58kDa fragment is present as a

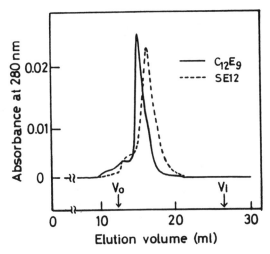

Figure 4. Elution profiles of the DIDS-labeled 58kDa fragment on a Shodex WS-803F column in 0.1 M Na_2HPO_4-NaH_2PO_4 buffer, pH 7.1, containing 0.1% $C_{12}E_9$(——) or 0.1% SE12 (----). (Reproduced with permission from Ref. 22. Copyright 1988 Elsevier Science Publishers.)

Table II. Molecular Parameters and Polypeptide Chain Molecular Weight of The 58kDa Fragment-Surfactant Complex*

Fragment-surfactant complex	Stokes radius of complex (A)	Amount of bound surfactant (g/g protein)	Sedimentation coefficient of complex $s_{20,w}$(S)	Polypeptide chain mole= cular weight ($\times 10^{-3}$)
$C_{12}E_9$ complex	75±3	1.0±0.1	4.4±0.2	126±13
SE12 complex	58±3	2.6±0.4	8.1±0.3	70±14

*values are an average of 3 or 4 determinations.
(Reproduced with permission from Ref. 22. Copyright 1988 Elsevier Science Publishers.)

monomer in SE12 solution rather than as a dimer, whereas the fragment exists as a dimer in $C_{12}E_9$ solution. Thus, SE12 might show higher efficiency than $C_{12}E_9$ disturbing hydrophobic intermolecular interactions between proteins.

Despite the disruption of the quarternary structure of the 58kDa fragment by SE12, this surfactant caused little change in the secondary structure of the polypeptide, as indicated by virtually identical profiles in the circular dichroic spectra of the 58kDa fragment in SE12 and $C_{12}E_9$ solutions (22). In addition, we measured the binding of BADS (a hydrophobic probe and a reversible anion transport inhibitor) to the 58kDa fragment in both SE12 and $C_{12}E_9$ solutions. It has been recognized (23) that the ability of BADS to bind to band 3 protein might depend on the maintenance of the protein functional structure for anion transport. The inhibitor was found to interact with the fragment in SE12 solution with slightly weaker affinity than in $C_{12}E_9$ solution, suggesting the conservation of functional structure in SE12 solution.

DISCUSSION

Among water-soluble proteins, the interaction of amphiphiles with bovine serum albumin had been receiving considerable attention, in relation with that protein functioning as a carrier for fatty acids in the circulatory system. A large body of experimental evidence has established (24,25) that different levels of affinity characterize these interactions. The hydrophobic patches or crevices on native bovine serum albumin provide a small number of discrete sites with high affinity for hydrophobic ligands. Following saturation of the high affinity sites, alkyl sulfates and sulfonates can bind to the protein in a cooperative mode. In this process, alkyl sulfates and sulfonates bind to sites which were previously buried and became exposed through conformational change. The cooperative mode of association characterizing those anionic surfactants, notably sodium dodecyl sulfate, is common to virtually all proteins.

Results presented here demonstrate that sucrose esters can occupy only the hydrophobic binding sites on native proteins, and they do not promote cooperative binding beyond that required for saturation of such sites. The small increase in the amount of bound SE10 at concentrations above the cmc, as observed in the case of bovine serum albumin, is due to the increase in monomer concentration. Thermodynamic analysis of the data shows the existence of associations between bovine serum albumin and SE10 monomers, rather than micelles. On the other hand, ovalbumin binds negligible amounts of sucrose esters, within experimental error. Comparison of the above results with existing data on protein-surfactant interactions unequivocally indicates that sucrose esters interact with proteins in a way similar to that of Triton X-100 (24,25), a well-characterized nonionic surfactant. Binding of SE10 to bovine serum albumin stabilized the native protein structure. The latter effect, however, should be viewed as a general phenomenon which occurs in the binding of amphiphiles to protein native sites. Hence, it is apparent that sucrose esters do not interact with proteins in some specific manner, as one would anticipate from the presence of the sucrose moiety; the latter can function as a protein stabilizer in its free state.

The mode of interaction between nonionic surfactants and hydrophobic proteins differs from that observed for hydrophilic proteins (24,25). In many studies with integral membrane proteins, the following has been observed: the protein in its isolated state accommodates a large number of surfactant molecules, covering areas of the protein normally embedded into the hydrocarbon core of the lipid bilayer. Such binding takes place cooperatively near the cmc and without any detectable change of protein conformation. A possible explanation for this phenomenon is that the hydrophobic region on the protein can act as nucleus for micelle formation ; the resulting complex resembles an approximately normal surfactant micelle into which the hydrophobic domain of the protein has been inserted, leaving the polar domain exposed to the solvent. It is not obvious whether the protein reacts with preformed micelles or rather interacts with monomeric surfactants in such a way as to govern the final structure of the complex. Regardless of the incomplete knowledge of the exact nature of their binding to hydrophobic proteins, it may be reasonable to expect that sucrose monoesters behave as other nonionic surfactants with hydrophobic proteins. The remarkably high binding number of SE12 to the 58kDa fragment observed here supports this hypothesis.

A characteristic of SE12 is that this surfactant has the ability to solubilize hydrophobic proteins which can hardly be extracted with other types of nonionic surfactants. An example of this is the solubilization of all components of β-adrenergic receptor protein from turkey erythrocyte membrane, resulting in a successful reconstitution (26). In addition to the efficient extractive power, it appears that SE12 can disturb hydrophobic interactions between nonpolar proteins. So far as the present data using the 58kDa fragment of band 3 protein are concerned, however, the ability is not strong enough to perturb the secondary structure of proteins.

ACKNOWLEDGMENTS

We thank Mr. Robert Houde for his critical reading of the manuscript. This work was supported in part by a grant awarded to SM by Mitsubishi Kasei Co.

LEGEND OF SYMBOLS

SEn, sucrose monoester of specific acid chain length; cmc, critical micelle concentration; Rs, Stokes radius; BADS, 4-benzamide-4'-aminostilbene 2,2'-disulfonate; DIDS, 4,4'-diisothiocyanostilbene 2,2'-disulfonate; $C_{12}E_9$, nonaethyleneglycol n-dodecylether.

LITERATURE CITED

1. Shibasaki, I. J. Food Safety 1982, 4, 35–58.
2. Marshall, D.L.; Bullerman, L.B. J. Food Prot. 1986, 49, 378–82.
3. Tsuchido, T.; Ahn, Y.H.; Takano, M. Appl. Environ. Microbiol. 1987, 53, 505–8.
4. Kawamura, S.; Murakami, M. Biochim. Biophys. Acta 1986, 870, 256–66.

5. Feder, D.; Im, M.J.; Klein, H.W.; Hekman, M.; Holzhofer, A.;
    Dees, C.; Levitzki, A.; Helmreich, E.J.M.; Pfeuer, T. EMBO
    J. 1986, 5, 1509-14.
6. Hekman, M.; Holzhofer, A.; Gierschik, P.; Im, M.J.; Jakobs,
    K.H.; Pfeuffer, T.; Helmreich, E.J.M. Eur. J. Biochem. 1987,
    169, 431-9.
7. Bourne, E.J.; Tiffin, A.I.; Weigel, H. J. Sci. Food Agric.
    1960, 11,101-9.
8. Ebeler, S.; Walker, C.E. J. Food Sci. 1984, 49, 380-4.
9. Ebeler, S.; Breyer, L.M.; Walker, C.E. J. Food Sci. 1986,
    51, 1276-9.
10. Matsunaga, A.; Kainuma, K. Starch 1986, 38, Nr. 1, S. 1-6.
11. Lee, J.C.; Timasheff, S.N. J. Biol. Chem. 1981, 256, 7193-
    201.
12. Nakashima, H.; Makino, S. J. Biochem. 1980, 88, 933-47.
13. Moriyama, R.; Kitahara, T.; Sasaki, T.; Makino, S. Arch.
    Biochem. Biophys. 1985, 243, 228-37.
14. Lotan, R.; Beattie, G.; Hubbel, W.; Nicolson, G.L. Biochemistry
    1977, 16, 1787-94.
15. Thompson, M.P.; Kiddy, C.A. J. Dairy. Sci. 1964, 47, 626-32.
16. Kinoshita, S.; Ohyama, S. Kogyo Kagaku Zasshi 1963, 66, 455-8.
17. Makino, S.; Ogimoto, S.; Koga, S. Agric. Biol. Chem. 1983,
    47, 319-26.
18. Tanford, C. The Hydrophobic Effect; John Wiley & Sons: New
    York, 1973; p 126.
19. Mercier, J.-C.; Grosclaude, F.; Ribadeau-Dumas, B. Eur. J.
    Biochem. 1971, 23, 41-51.
20. Yang, J.T.; Foster, J.F. J. Amer. Chem. Soc. 1953, 75, 5560-7.
21. Reynolds, J.A.; Herbert, S.; Polet, H.; Steinhardt, J.
    Biochemistry 1967, 6, 937-47.
22. Tomida, M.; Kondo, Y.; Moriyama, R.; Machida, H.; Makino, S.
    Biochim. Biophys. Acta 1988, 943, 493-500.
23. Moriyama, R.; Makino, S. Biochim. Biophys. Acta 1985, 832,
    135-41.
24. Helenius, A; Simons, K. Biochim. Biophys. Acta 1975, 415,
    29-79.
25. Makino, S. in Adv. Biophys.; Kotani, M., Ed.; Japan Scientific
    Society: Tokyo, 1979; Vol. 12, p 131.
26. Hekman, M.; Feder, D.; Keenen, A.K.; Gal. A.; Klein, H.W.;
    Pfeuffer, T.; Levitzki, A.; Helmreich, E.J.M. EMBO J. 1984,
    3, 3339-45.

RECEIVED August 7, 1990

Chapter 15

Importance of Hydrophobicity of Proteins in Food Emulsions

E. Li-Chan and S. Nakai

Department of Food Science, University of British Columbia,
6650 NW Marine Drive, Vancouver, British Columbia V6T 1W5, Canada

The amphiphilic nature of proteins is important for
their function as emulsifiers. Of particular relevance
is the hydrophobic nature at the molecular surface, in
conjunction with steric effects or flexibility which
allow exposure of previously buried groups during
emulsification. Various methods have been proposed to
quantitate surface hydrophobicity of food proteins for
elucidation of their emulsifying properties, including
the use of fluorescence probes to investigate dilute
protein solutions. This chapter discusses application
of Raman spectroscopy to study hydrophobic interactions
of proteins in concentrated solutions or gelled states,
as well as in interactions with lipids in emulsions.

Quantitative structure-activity relationship (QSAR) techniques use
molecular structure and physical property data to make predictions
about activity and reactivity of compounds. Hydrophobicity,
topological descriptors, electronic descriptors and steric effects
are common structure/property descriptors used in QSAR in
environmental studies and in pharmaceutical research (1,2). The
importance of hydrophobic, steric and charge parameters also extends
to QSAR for elucidating functionality of proteins in food systems,
including emulsifying properties (3,4). In particular, the
importance of the amphiphilic nature of proteins in their role as
emulsifiers has been widely recognized. Many methods have been
proposed to measure the hydrophobicity of proteins which may be
important in their emulsifying functions. At the same time, many
researchers have focussed on processes during diffusion, adsorption
and rearrangment of protein molecules at surfaces or interfaces. In
most cases, dilute protein solutions and model hydrocarbons or
triglycerides have been used in these studies.
    In order to obtain information which can be truly useful in
elucidating QSAR in food emulsions, more effort should be focussed on

0097–6156/91/0448–0193$06.00/0

the complex world of real food systems rather than the model system (5,6). For example, what is the concentration of protein commonly occurring in the food emulsion? Is the process of adsorption at the interface more likely to be diffusion- or turbulence-controlled? Can complex animal and vegetable fats and oils be used to study protein-lipid interactions?

In this chapter, the importance of hydrophobic interactions of proteins in food emulsions is discussed by considering some of these issues. The role of steric effects and molecular flexibility on exposure of hydrophobic groups at the surface of the protein molecules which can participate in emulsification is emphasized. A review of some empirical methods currently used to measure protein hydrophobicity is followed by recent research in our laboratory applying laser Raman spectroscopy to study hydrophobic interactions and steric effects of proteins in solution, gels and emulsions.

Molecular Structure, Hydrophobicity and Emulsifying Properties

The molecular processes which occur during emulsification in food systems have been the subject of several recent reviews and books (eg. 7-10), and consist of a number of stages that include migration to the oil-water interface, adsorption at the interface, conformational re-arrangement and formation of multiple layers. Desorption or reversibility of the adsorption process is also possible, especially in the case of mixed emulsifiers. For many years, it was accepted that the solubility of a protein was the primary determinant in its emulsifying properties (11). However, in the last decade, key roles have been assigned to both backbone flexibility and proper hydrophobic-hydrophilic balance of the protein molecule. Solubility facilitates diffusion of molecules to the surface or interface. However, whether or not this is a critical factor under the turbulent conditions of high shear and energy input which exist during emulsification is unclear. In any case, once at the interface, the ability of the protein to interact with the oil or with other protein molecules to form an interfacial layer depends on its flexibility and accessibility of surface exposed groups for interactions, especially through hydrophobic interactions. Charge effects play an additional important role since electrostatic repulsions can hinder intermolecular interactions at the interface.

β-casein from cow's milk is an example of a food protein which possesses both high flexibility and amphiphilicity and is recognized as a superior emulsifier. However, many food proteins are globular in nature and are often limited in both molecular flexibility as well as hydrophobic groups which are exposed on the surface of the protein and available to participate in its functionality. For example, lysozyme from hen egg white has a compact globular structure which is stabilized by four disulfide cross-links. Lysozyme exhibits poor emulsifying and foaming properties, and is resistant to irreversible denaturation by heating up to $75^{\circ}C$. The often-cited work of Graham and Phillips (12-14), which compared the behavior between β-casein, lysozyme and bovine serum albumin at air-water and "oil"-water or toluene-water interfaces, is an excellent demonstration of the effects of molecular structure on their properties. However, as suggested by Mangino (6), applicability of their conclusions to food emulsions is questionable due to the low protein concentrations at which the differences in behavior were observed.

Surface or Exposed Hydrophobicity. The concept of hydrophile-lipophile balance or HLB in expressing the relative proportions of hydrophobic (or lipophilic or nonpolar) versus hydrophilic (or polar) moieties in low molecular weight compounds has been a useful aid in the selection of suitable surface active agents for formation of oil-in-water or water-in-oil emulsions. In the case of proteins, a value analogous to HLB can be obtained by summing up or taking the average of the hydrophobicity values of the constituent amino acid residues of the protein. The calculation of total or average hydrophobicity by the method of Bigelow (15) is an example of this. However, such values do not consider the effect of protein structure on the extent of exposure of residues. Although it is generally agreed that charged amino acid residues are located preferentially at the surface of globular proteins while nonpolar or hydrophobic residues are buried in the interior of the molecule, analysis of the three-dimensional structure of proteins by techniques such as X-ray crystallography has indicated the presence of hydrophobic patches on the surface of proteins. It is therefore likely that the groups which can participate in protein functionality, such as emulsification, are those amino acid residues which are located on the surface of the native protein molecules or become exposed during processing, such as heating or homogenization or whipping.

The observation that proteins can decrease interfacial tension at an oil-water interface has been suggested to be a factor in their ability to act as emulsifiers. Keshavarz and Nakai (16) first showed that interfacial tension values at 0.2% protein solution/corn oil interfaces were negatively correlated with hydrophobicity of proteins determined by either retention volume on hydrophobic interaction chromatography or partition coefficient between phases of differing polarity. No correlation was obtained between these "effective" hydrophobicity values and average hydrophobicity values of the proteins calculated by Bigelow's method. The correlation between emulsifying activity and "surface" hydrophobicity of proteins was subsequently demonstrated by Kato and Nakai (17), who measured surface hydrophobicity by the increase in fluorescence intensity upon binding of hydrophobic probes to the protein solutions.

Molecular flexibility. Based on the adsorption behavior of linear polymers at solid surfaces, it has been proposed that proteins undergo conformational rearrangement at the interface resulting in segments referred to as trains, loops and tails. However, the validity of this model has been controversial.

In cases where adsorption at the interface occurs from concentrated bulk solution, it has been suggested that a mixture of native and denatured molecules exists, and the physical state of the adsorbed film may resemble that of a gel (18). Change in the conformation of whey proteins adsorbed on emulsified fat globules was suggested by their greater susceptibility to protease digestion than the whey proteins in aqueous solution (19). No change in the secondary structure of lysozyme upon adsorption was observed by circular dichroism (20), but it was not clear if any tertiary structural changes had occurred or whether such changes were restricted by the disulfide crosslinks stabilizing this protein. Despite the lack of consensus on the actual structural changes which occur upon and after adsorption, it is probably agreed that a certain degree of flexibility is required to promote surface or interfacial activity. Amphiphilic proteins may be encouraged to adsorb at an interface by partial denaturation such as that brought about by mild conditions of heating. Kato et al. (21-23) reported that heat treatment which was controlled so as not to produce coagulation resulted in increased surface hydrophobicity of the heated proteins, and that these increased hydrophobicity values were linearly related to improved emulsifying properties. On the other hand, when intramolecular crosslinks were chemically introduced to a protein such as bovine serum albumin, surface active properties were dramatically impaired (24).

Effect of Disulfide Bond Reduction. Disulfide bonds stabilize the tertiary structure of many globular proteins such as lysozyme, bovine serum albumin and soy glycinin. Treatment of these proteins with a reducing agent such as dithiothreitol (DTT) was reported to improve their surface active properties, including foaming and emulsifying properties, as well as to induce gelation in some cases (25-29).
In the case of soy glycinin, reduction of 13 of the 20 disulfide bonds by treatment with 5 mM DTT in the presence of 8 M urea, followed by blocking of sulfhydryl groups with iodoacetamide, resulted in dramatic increases in surface hydrophobicity apparently through cleavage of intermolecular bonds linking the subunits; reduction of all 20 disulfide bonds, both intra- and intermolecular, was achieved using 10 mM DTT. The molecular conformational changes which enhanced molecular hydrophobicity and increased viscosity were suggested to be responsible for the significant improvement in surface active properties of the reduced proteins (25,26). However, the harsh conditions during the reducing treatment as well as possible changes in the protein surface character by use of a blocking agent make it difficult to conclude that the improvements in surface active properties were due to enhanced molecular flexibility arising from reduction per se.
When lysozyme solutions (0.5% in 50 mM NaCl at pH 7) were treated using much milder conditions, i.e. with 0.7 mM DTT at either 4°C or room temperature for up to 24 hours, only slight increase was observed both in the sulfhydryl content and in surface hydrophobicity determined by fluorescence probes (29). Nevertheless, a significant increase in the emulsifying activity index was observed in the DTT-treated lysozyme compared to the control. It was suggested that the mechanical energy of the emulsification process itself was sufficient to enable unfolding of the partially reduced lysozyme. DTT-treated lysozyme subjected to the emulsification process in the absence of any oil exhibited a significantly higher surface hydrophobicity value compared to similarly emulsified lysozyme solution. Upon heating of 0.5% solutions of the DTT-treated lysoyzme either at 80°C for 12 minutes or at 37°C for 24 hours, up to 3 of the 4 disulfide bonds of lysozyme were reduced, and significant increases in surface hydrophobicity, emulsifying activity

and stability as well as in turbidity of the solutions were noted. These results demonstrate that changes in molecular flexibility allowing exposure of hydrophobic groups can be brought about by the cleavage of disulfide bonds using low concentrations of reducing agent combined with energy input in the form of the emulsification process itself or by heating. In other words, the potential for hydrophobic interactions at the interface is the key factor, rather than highly hydrophobic characteristics of the protein molecules in bulk solution as the latter can lead to formation of aggregates and may in fact hinder adsorption at the interface.

Effect of Protein Concentration in the Bulk Solution. The effects of concentration of protein molecules in the bulk solution to the interfacial protein concentration and characteristics of adsorption and multilayer formation have been the subject of much controversy. Under quiescent conditions of adsorption and at low bulk protein concentration, the adsorbed layer has been proposed to resemble a two-dimensional gas with unfolded molecules; at intermediate concentrations, the surface layer becomes compressed and resembles a condensed liquid film, while at still higher concentrations, the adsorbed layer may exhibit viscoelastic or solid-like properties (5). However, adsorption under quiescent conditions can differ from that in an emulsion, and it has been suggested that the much higher surface-to-volume ratio in the latter may result in the significant depletion of protein from the bulk solution, at low initial bulk concentrations (30).

Many studies have indicated that emulsifying capacity or activity expressed per unit weight of protein decreases as a function of increasing bulk concentration of protein, and differences in emulsifying properties between proteins have been reported to be negligible at high protein concentration. Halling (30) indicated that the differences in emulsifying properties exhibited with respect to protein concentration are actually a reflection of differing soluble protein content. However, as illustrated in Figure 1, recent work in our laboratory (Lee, G. University of British Columbia, B.Sc. thesis, 1989) has demonstrated that there are inherent differences between soluble proteins in their emulsifying activity behavior as a function of protein concentration. Viscosity of the continuous phase has been cited as a parameter affecting emulsion stability and may be altered by protein concentration. However, the relationship is not a simple one; no significant correlation could be found between the turbidity and viscosity of emulsions of various proteins at different initial bulk concentration (Figures 1a and 1b). Although increased viscosity may impart greater transient droplet stability during the emulsification process and thus reduce re-coalescence, the viscosity of the continuous phase has a small effect on final droplet size, and in fact highly viscous aqueous phases containing cellulose derivatives were observed to yield larger droplet sizes (5). Dickinson and Stainsby (7) concluded that empirical studies have not revealed any general rules about the relationship between protein concentration and emulsion stability.

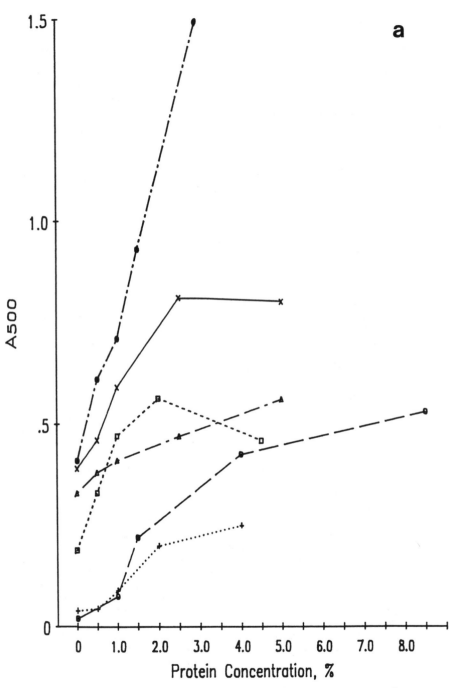

Figure 1. (a) Emulsifying activity (expressed as A_{500}) as a function of initial bulk solution concentration. (X, bovine serum albumin; O, lysozyme; □, casein; Δ, B-lactoglobulin; +, ovalbumin; ●, gelatin)

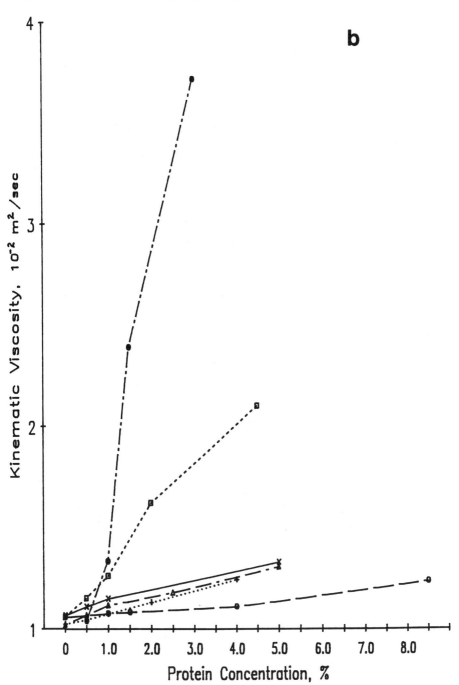

Figure 1. (b) Kinematic viscosity of various soluble proteins as a function of initial bulk solution

Comparison of Current Methods for Protein Hydrophobicity Studies

Methods for quantitative estimation of protein hydrophobicity values have been recently reviewed (3 and Li-Chan, E. In Encyclopedia of Food Science and Technology; Hui, Y. H., Ed.; Wiley-Interscience: New York, 1990; in press). Generally they may be categorized into (1) various algorithms for computation of hydrophobicity values or profiles, using hydrophobicity scales of the individual amino acids and data of the amino acid composition or primary sequence of the protein; (2) partition methods, including relative solubility in polar and nonpolar solvents or relative retention behavior on reverse phase or hydrophobic interaction chromatography; (3) binding methods, including the binding of aliphatic and aromatic hydrocarbons, sodium dodecyl sulfate, simple triglycerides and corn oil; and (4) spectroscopic methods, including intrinsic fluorescence, derivative spectroscopy and use of fluorescence probes.

Considerable variations in the values obtained for proteins by different methods have been reported, and caution is warranted whenever the method involves use of nonpolar organic solvents which may alter the native structure and thus the surface hydrophobic characteristics of the protein. Spectroscopic methods are usually non-destructive and less likely to induce conformational changes in the protein molecule. However, methods such as ultraviolet absorbance, derivative spectrophotometry, or intrinsic fluorescence, usually only give information about the aromatic chromophores of the protein. Algorithms for quantitation of protein hydrophobicity, while useful for certain proteins, are limited in universality of application, due to the need for considerable information on sequence or structure of the protein, assumption of homologous behaviour of residues of the protein under investigation to those in the database upon which the algorithms were formulated, and inability to evaluate hydrophobicity of complex food systems containing several protein species, whose interactions may change during food processing.

Extrinsic fluorescence probes have become widely used to probe hydrophobic sites of proteins, probably due to the rapidity and simplicity of the methodology involved. The most commonly used probes in the study of food proteins have been the aromatic hydrophobic probe, 1-anilinonaphthalene-8-sulfonic acid or ANS, and the aliphatic hydrophobic probe, cis-parinaric acid or CPA. The latter may be particularly advantageous as its structure is analogous to fatty acids and thus it may probe for sites on the protein molecule which can bind food lipids. During this past decade, many reports have appeared in the literature on the use of these fluorescence probes to investigate hydrophobic interactions of food protein systems and how the surface hydrophobic groups may be changed by processing (3). However, despite the widespread use and potential fundamental information which can be obtained, two limitations may restrict the interpretation of results obtained by these fluorescence probes : (1) Both ANS and CPA contain anionic moieties which may cast doubt on whether their probing effect is solely for hydrophobic sites; the effect of rigidity of the environment rather than hydrophobicity on fluorescence of ANS has also been cautioned; (2) low protein concentrations (typically 0.005-0.05%) are used for the measurement of fluorescence of the bound probe, which is in contrast to the much higher concentrations which would typically be encountered in real food systems such as emulsions. Thus other methods to quantitate hydrophobic interactions are being sought to supplement the information which has been acquired through fluorescence probes. These new methods should be applicable at higher protein concentrations, and without using extrinsic probes which may perturb the native protein structure. Two approaches currently being investigated in our laboratory are the use of proton magnetic resonance to determine extent of exposure of aliphatic and aromatic side chains of proteins under different conditions, and application of Raman spectroscopy (31). The remainder of this chapter will present some of the results which we have recently obtained using the latter technique.

Raman Spectroscopy

Raman spectroscopy is a light scattering technique which can give detailed information on the vibrational motions of atoms in molecules. In essence, the technique based on Raman scattering measures shifts in the wavelength of the exciting laser beam which arise from the inelastic collisions between the sample molecules and the photons or particles of light composing the laser beam. Because the intensity and frequency of the molecular vibrations are sensitive to chemical changes and the environment of the atoms, the Raman spectrum can be used as a monitor of molecular chemistry. Table I shows typical assignments of amino acid residue side chain vibrations which correspond to peaks in different wavenumber shift regions (cm^{-1}) of the protein Raman spectrum. Additional information can be obtained from vibrations arising from the peptide backbone; analysis of the amide I (1640-1680 cm^{-1}) and amide III (1230-1300 cm^{-1}) regions in particular can yield useful data on the secondary structure of the protein (32-34).

   Since the early research on peptides and proteins in the 1960's, Raman spectroscopic applications in the study of biological molecules including proteins have been rapidly expanding, as exemplified in some recent monographs (35-39). Some distinct advantages in the application of Raman over other spectroscopic methods are that it can be applied to aqueous solutions of molecules as well as nonaqueous liquids, and has been successfully used to study molecules in fibers, films, powders, gels and crystalline solids of biopolymers. Thus, it has great potential to study changes in the protein molecule in the dry versus hydrated or solution state, in solvents of differing polarity, or after coagulation or gelation, as well as to monitor interactions with other molecules such as lipids and carbohydrates. Painter (40) presented an excellent overview of some applications of Raman spectroscopy to the characterization of food, and noted that this technique has a clear but as yet unrealized potential for characterizing the individual components of food systems.

   The pioneering work of Professor Richard Collins Lord yielded the first interpretable laser Raman spectrum of a native protein, lysozyme in aqueous solution. In a series of papers published in the early 1970's, Lord and his co-workers studied the Raman spectrum of native lysozyme as well as the changes observed by extensive

Table I. Assignments of typical amino acid side chain vibrations
in Raman spectrum of proteins (compiled from references 35-38).

cm^{-1}	vibration assignment/interpretation
2700-3300	CH stretch ⎫
1465 ± 20	CH_2 bend ⎬ aliphatic
1450 ± 20	CH_3 antisymmetric bend ⎭ side chains
1399	ser, thr OH (weak)
1725-1700	asp, glu COOH, un-ionized C=O stretch
1425	asp, glu COOH, ionized C=O stretch
1650(asn),1615 (gln)	asn/gln amide
1640,1600	lys NH_3^+
1491	his imidazole (probe ionization state)
1605,1585,1207,1006,622	phe
1600,1590,850,830	tyr (850/830 is environment sensitive)
1582,1553,1363,1014,879	trp (1360 and 880 cm^{-1} intensity are
761,577,544	especially sensitive to environment)
540	trans-gauche-trans ⎫
525	gauche-gauche-trans ⎬ cystine SS
510	gauche-gauche-gauche ⎭
745-700	trans C-S (met, cys, cys/2)
670-630	gauche C-S (met, cys, cys/2)

denaturation using heat, complete reduction of disulfide bonds or with LiBr or SDS (41-43). Based on this work, they concluded that the denatured protein exists in random-coil conformations, which may differ depending on the various denaturing agents. Recently, we have used Raman spectroscopy to study changes in lysozyme in concentrated (10-20%) solutions after treatment by heat and/or mild DTT reduction. The objective of this work was to investigate changes brought about by mild reduction (in contrast to complete reduction and extensive denaturation reported by Lord and co-workers), which were able to confer improved gelling, foaming and emulsifying properties on lysozyme, and to compare these changes with those previously demonstrated at lower concentration (29). In addition, application of Raman spectroscopy to study interactions of protein with corn oil is also demonsrated.

Experimental. Raman spectra were recorded on a JASCO model NR-1100 laser Raman spectrophotometer with excitation from the 488 nm line of a Spectra-Physics Model 168B argon ion laser. The spectra of 10 or 20% (w/w) solutions of lysozyme (Sigma L6876) in water (final pH 5.75) or in deuterium oxide (final apparent pD 5.85), corn oil (commercial grade Mazola brand), or their emulsions in hematocrit capillary tubes were measured at ambient temperature under the following conditions: laser power 200 mW, slit height 2 mm; spectral resolution of 5.0 cm^{-1} at 19,000 cm^{-1}, sampling speed 120 cm^{-1}/min with data taken every cm^{-1}, 6-10 scans per sample. For samples treated with DTT (DL-dithiothreitol, Sigma D0632) and heating, solutions of the samples were placed in the capillary tubes prior to heat-induced gelation. Background correction, normalization, smoothing and difference spectrum computation of the recorded spectra were performed with the NR-1100 data station. Further details of the procedures are reported elsewhere (Li-Chan and Nakai, 1990; manuscripts in preparation).

Raman Spectra of Lysozyme Solution and Heat Induced Gels. Figures 2 and 3 show the Raman spectra of 20% lysozyme solutions and the gels formed by heating at 100°C for 5 minutes. The changes brought about by this short heat treatment at high protein concentration are generally similar to those reported when 7% solutions were heated at 100°C for 2 hours (42). Some of the more evident differences arising from heating include the following: (1) the relative intensity of the 510 and 525 cm^{-1} peaks in the disulfide (SS) stretching region changes from 3:1 to 1:1 ratio, which corresponds to a change from an all gauche to a gauche-gauche-trans conformation, respectively; (2) decrease in intensity of the peaks at 760, 880 and 1360 cm^{-1}, which indicate increased exposure of tryptophan residues to the aqueous environment; (3) shifts in the centre of the amide III region from 1257 to 1245 cm^{-1}, and in the amide I region from 1661 to a doublet at 1660 and 1674 cm^{-1}, which are related to a decrease in α-helical structure and increases in β-sheet and random coil structure; (4) increase in the intensity of the CH stretching vibration at 2938 cm^{-1} relative to the broad water line at 3230 cm^{-1} which suggest increased exposure of aliphatic side chains as well as decreased mobility of the water molecules after heat-induced gelation. These changes correlate well with the increases in hydrophobicity after heating measured by methods such as fluorescence

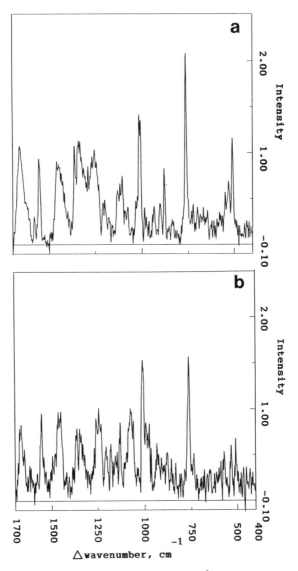

Figure 2. Raman spectra (400-1700 cm^{-1} shift region) of
20% lysozyme (a) solution and (b) gel formed after heating at
100°C for 5 minutes.
(Spectra were normalized to the intensity of the H-C-H
deformation mode at 1455 cm^{-1}, after baseline correction, as
recommended in reference 42).

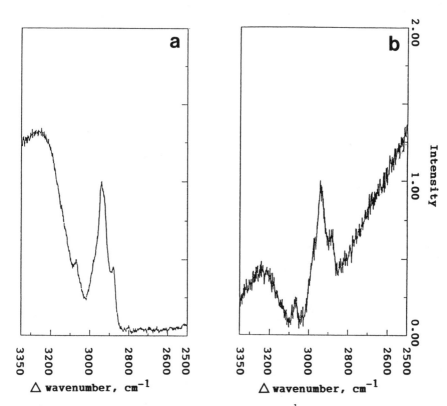

Figure 3. Raman spectra (2500-3350 cm^{-1} shift region) of 20% lysozyme (a) solution and (b) gel formed after heating at 100°C for 5 minutes.
(Spectra were normalized to the intensity of the C-H stretching band at 2940 cm^{-1}).

probes (29), and with the suggestion that heat-induced gelation does
not always result in complete unfolding to a random coil but in fact,
there is often an increase in β-sheet structure (44). Furthermore,
although chemical analysis using Edman's reagent did not show any
significant changes in sulfhydryl group content after heating, it is
interesting to note the capability of Raman spectroscopy to detect
changes in the disulfide bond conformation in the gel.

Effect of DTT and/or Heating on Raman Spectra of Lysozyme.   The
Raman spectra of unheated 10-20% lysozyme solutions treated with 10
mM DTT did not appear to differ from that of lysozyme, which is
consistent with the small changes in both sulfhydryl content and
hydrophobicity measured by fluorescence probes reported previously.
Firm opaque white gels were formed when lysozyme solutions at ≥5%
concentration were heated at either low (37-40°C) or high
(75-80°C) temperature in the presence of low concentrations of DTT
(29). The physical changes manifested in gel formation were
accompanied by marked changes in the Raman spectra. Figure 4 shows
the Raman difference spectra of deuterated lysozyme with DTT and/or
heat treatment. In the presence of DTT, both low and high
temperature treatment resulted in the exposure of aromatic residues
(decrease in peak intensity at 760, 1005-1010, 1200 and 1555 cm^{-1}),
decrease in α-helical content (930-950 cm^{-1}), increase in
antiparallel β-sheet structure (980-990 cm^{-1}), changes in the
disulfide stretching conformations (505-530 cm^{-1}) (35) and increase
in peptide backbone vibrations (C-C and C-N at 1050-1150 cm^{-1})
(37). The changes were more marked after heating at 75°C, and
additional changes at the higher temperature were noted in the H-C-H
bending vibrations at 1445 cm^{-1} and the tryptophanyl side chain
vibrations at 1330-1340 cm^{-1}.
        In general, the changes noted in the Raman spectra of
concentrated solutions and gels confirm those previously observed for
increased surface hydrophobicity determined on dilute solutions.
However, additional information on changes in secondary structure and
more detailed analyses of the changes in aromatic and aliphatic
residues are possible. Furthermore, a distinct influence of protein
concentration was observed for effect of DTT and heat treatment on
resulting sulfhydryl (SH) content of the protein. Whereas 3.4 and
6.0 moles SH/mole protein resulted after heating 0.5% lysozyme
solutions containing 0.7 mM DTT at 37°C and 80°C, respectively,
only 0.9 and 0.5 mole SH/mole, respectively, resulted by heating 10%
lysozyme solutions in the presence of 10 mM DTT. A peak
corresponding to free sulfhydryl (2570-2580 cm^{-1}) (35) could not be
readily detected in the Raman spectrum of the latter samples, but was
detected in the spectrum of 10% lysozyme heated in the presence of
100 mM DTT, in which case 6 moles SH/mole were determined by Ellman's
reagent.

Raman Spectra of Interactions of Protein with Corn oil. Figure 5
shows the Raman spectra of DTT-treated lysozyme, corn oil and
emulsions formed by either vortexing or sonication of protein
solution with oil (oil phase volume or Ø of 0.25). The Raman spectra
in this region corresponds to CH vibrational bands of both protein
and oil components, as shown in Table II (adapted from 35).

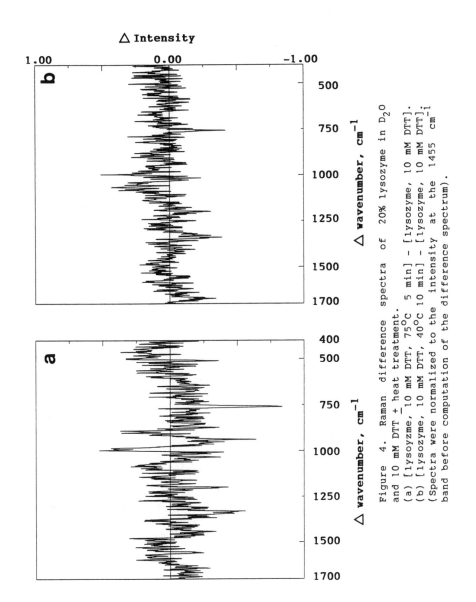

Figure 4. Raman difference spectra of 20% lysozyme in D_2O and 10 mM DTT + heat treatment. (a) [lysoyzme, 10 mM DTT, 75°C 5 min] − [lysozyme, 10 mM DTT]. (b) [lysozyme, 10 mM DTT, 40°C 10 min] − [lysozyme, 10 mM DTT]. (Spectra were normalized to the intensity at the 1455 cm^{-1} band before computation of the difference spectrum).

Table II.   C-H vibrational bands from protein and lipid components

Vibration	Wavenumber, cm^{-1}	
	Protein	Lipid
CH_2 symmetric	2900	2850
CH_2 asymmetric	2940	2885
CH_3 asymmetric	2980	2930

The spectrum of corn oil shown in Figure 5 resembles that of oleic and linoleic acids in the liquid form reported by Verma and Wallach (45), which may be expected from the fatty acid composition reported for this oil.   In addition to the vibrational assignments noted in Table II, corn oil also exhibits a peak at 3012 cm^{-1} which corresponds to CH=CH vibration of unsaturated fatty acids. The spectra of the emulsions formed using DTT-lysozyme appear to be a composite of the protein and corn oil spectra.   Similar spectra were observed for sonicated emulsions formed using lysozyme in the absence or presence of DTT.   However, the emulsion formed by simple vortex action using lysozyme without DTT was unstable, separating into two layers within minutes.

Figure 6 shows the Raman difference spectra of the sonicated and vortexed mixtures after subtraction of the corn oil and respective protein spectra from the emulsion spectra.   These difference spectra showed changes in the vibrational parameters of the molecules which may represent the interactions of the oil and protein components resulting from the emulsification process.   Small negative peaks at 2930 and 2980 cm^{-1} observed in the difference spectrum of vortexed lysozyme emulsion may be due to changes in Raman scattering arising from CH_3 asymmetric vibrations of both protein and lipid components. Greater extent of protein-lipid interactions are suggested in the difference spectrum for vortexed DTT-lysozyme emulsion.   The sharp negative peak at 2848 cm^{-1} and smaller negative peak at 3005 cm^{-1} may reflect changes in Raman scattering attributed to CH_2 and CH=CH vibrations of the oil, while the increased intensity at 2940 and 2980 cm^{-1} may be due to changes attributed to the protein asymmetric vibrations.   Increase in Raman active CH stretching of proteins near 2930 cm^{-1} was suggested to reflect exposure of apolar side chains to an aqueous milieu (46).

Much greater changes are seen in the difference spectra of the sonicated emulsions.   The increase in intensity at 2900 and 2970-2980 cm^{-1} and decrease at 2930-2940 cm^{-1} regions are likely due to changes in environment of the protein aliphatic chains.   In addition, there appear to be shifts to higher wavenumber for some of the peaks in the emulsion spectrum compared to the individual spectra of the protein and oil eg. negative peaks at 2849, 2875, 2930 and 3005, compared to positive peaks at 2858, 2895, 2970 and 3015-3020 cm^{-1}).   Verma and Wallach (46) reported that the CH stretch bands of methanol, dimethylsulfoxide and t-butanol shifted to higher frequency when the pure solvents were mixed with water, ie. in a more polar environment.   They also noted shifts in the wavenumbers of peaks of fatty acids in the solid versus liquid form, and reported a more complex pattern especially below 2880 cm^{-1} for unsaturated C_{18} fatty acids in the solid form, which they attributed to

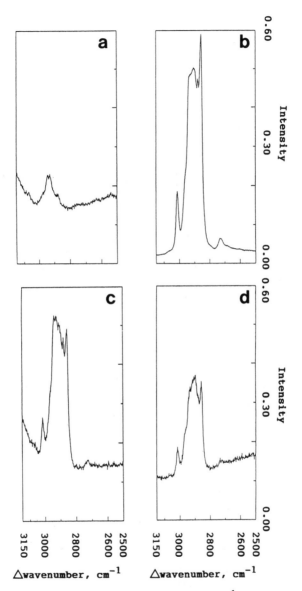

Figure 5. Raman spectra in the 2500-3150 cm^{-1} shift region of (a) 10% lysozyme + 10 mM DTT; (b) corn oil; and their emulsions (ϕ=0.25) formed by (c) vortex action (30 seconds) or (d) sonication (15 watts, 10 min at 50% pulse rate, in ice bath).

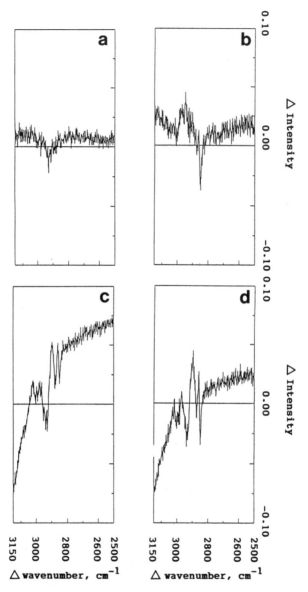

Figure 6. Raman difference spectra of (a) vortexed lysozyme,
(b) vortexed DTT-lysozyme, (c) sonicated lysozyme and (d)
sonicated DTT-lysozyme and corn oil emulsions, prepared as
described in the legend to Figure 5. Difference spectra were
computed on the JASCO NR-1100 data station, using standard
spectral wavenumbers of 2940 and 3015 cm^{-1} for subtraction of
protein and oil spectra, respectively, from the emulsion
spectra.

non-equivalence in environment of the CH_2 residues which is accentuated in the solid state. The complex pattern seen in the difference spectrum of the sonicated emulsions may thus be a reflection of interactions which lead to immobilization of the fatty acid chains. An additional feature which can be noted in these difference spectra is the highly negative slope at wavenumber above 3000 cm^{-1} corresponding to the water band, which may be a result of changes in its Raman scattering coefficient and may suggest an immobilization of water molecules in these emulsions.

Conclusions

Molecular flexibility and the potential for hydrophobic interactions of proteins with other protein or lipid molecules at the oil-water interface are key structural factors in the ability of proteins to act as emulsifiers. While fluorescence probes can be used to study hydrophobic interactions of proteins in dilute solutions, Raman spectroscopy may be applied to investigate the interactions of proteins in more concentrated solutions, gels and in emulsions.

Literature Cited
1. Borman, S. Chemical & Engineering News 1990, 68(8), 20-23.
2. Hansch, C.; Clayton, J.M. J. Pharm. Sci. 1973, 62, 1-21.
3. Nakai, S.; Li-Chan, E. Hydrophobic Interactions in Food Systems; CRC: Boca Raton, Florida, 1988.
4. Nakai, S. J. Agric. Food Chem. 1983, 31, 676-683.
5. Darling, D.F.; Birkett, R.J. In Food Emulsions and Foams; Dickinson, E., Ed.; Royal Society of Chemistry: London, 1987; Chapter 1.
6. Mangino, M.E. In Food Proteins; Kinsella, J.E.; Soucie, W.G., Eds.; Amer. Oil Chem. Soc.: Champaign, Il, 1989; Chapter 9.
7. Dickinson, E.; Stainsby, G. Colloids in Food; Applied Science:London, 1982.
8. Dickinson, E.; Stainsby, G., Eds. Advances in Food Emulsions and Foods; Elsevier Applied Science: London and NY, 1988.
9. Dickinson, E., Ed. Food Emulsions and Foams; Royal Society of Chemistry:London, 1987.
10. Becher, P. Encyclopaedia of Emulsion Technology, Volumes 1 and 2; Marcel Dekker: New York; 1983.
11. Kinsella, J.E. Crit. Rev. Food Sci. Nutr. 1976, 7, 219-280.
12. Graham, D.E.; Phillips, M.C. In Foams; Akers, R.J., Ed.; Academic: New York; 1976, pp. 237-255.
13. Graham, D.E.; Phillips, M.C. J. Colloid Interfac. Sci. 1979, 70, 403-414; 415-426; 427-439.
14. Graham, D.E.; Phillips, M.C. J. Colloid Interfac. Sci. 1980, 76, 227-339; 240-249.
15. Bigelow, C.C. J. Theoret. Biol. 1967, 16, 187-211.
16. Keshavarz, E.; Nakai, S. Biochim. Biophys. Acta 1979, 576, 269-279.
17. Kato, A.; Nakai, S. Biochim. Biophys. Acta 1980, 624, 13-20.
18. Dickinson, E.; Murray, B.S.; Stainsby, G. In Advances in Food Emulsions and Foams; Dickinson, E.; Stainsby, G., Eds.; Elsevier Applied Science:London, 1988; Chapter 4.
19. Shimizu, M.; Kamiya, T.; Yamauchi, K. Agric.Biol. Chem. 1981, 45, 2491-2496.

20. Izmailova, V.N.; Yampol'skaya, B.P.; Lapina, G.P.; Sorokin, M.M. Colloid J. USSR 1982, 44, 195. Cited in Reference 18.
21. Kato, A.; Tsutsui, N.; Matsudomi, N.; Kobayashi, K; Nakai, S. Agric. Biol. Chem. 1981, 45, 2755-2760.
22. Kato, A.; Osako, Y.; Matsudomi, N.; Kobayashi, K. Agric. Biol. Chem. 1983, 47, 33-37.
23. Kato, A.; Komatsu, K.; Fujimoto, K.; Kobayashi, K. J. Agric. Food Chem. 1985, 33, 931-934.
24. Kato, A.; Yamaoka, H.; Matsudomi, N.; Kobayashi, K. J. Agric.Food Chem. 1986, 34, 370-372.
25. Kim, S.H.; Kinsella, J.E. J. Agric. Food Chem. 1986, 34, 623-7.
26. Kim, S.H.; Kinsella, J.E. J. Food Sci. 1987, 52, 128-131.
27. Nakamura, R.; Mizutani, R.; Yano, M.; Hayakawa, S. J. Agric. Food Chem. 1988, 36, 729-732.
28. Haque, Z.; Kinsella, J.E. J. Food Sci. 1988, 53, 416-420.
29. Li-Chan, E.; Nakai, S. In Food Proteins; Kinsella, J.E.; Soucie, W.G., Eds.; Amer. Oil Chem. Soc.:Champaign, Il, 1989; Chapter 14.
30. Halling, P.J. CRC Crit.Rev.Food Sci. Nutr. 1981, 15, 155-203.
31. Nakai, S.; Li-Chan, E. International Chemical Congress of Pacific Basin Societies; Honolulu, 1989; Abstract 162.
32. Przybycien, T.M.; Bailey, J.E. Biochim. Biophys. Acta 1989, 995, 231-245.
33. Byler, D.M.; Farrell, H.M. Jr.; Susi, H. J. Dairy Sci. 1988, 71, 2622-2629.
34. Berjot, M.; Marx, J.; Alix, J.P. J. Raman Spec. 1987, 18, 289-300.
35. Tu, A.T. In Advances in Spectroscopy Vol. 13; Clark, R.J.H.; Hester, R.E., Eds.; Wiley: New York; 1986; Chapter 2.
36. Carey, P.R. Biochemical Applications of Raman and Resonance Raman Spectroscopies; Academic Press: New York; 1982.
37. Alix, A.J.P.; Bernard, L.; Manfait, M., Eds. Spectroscopy of Biological Molecules; Wiley: New York; 1985.
38. Spiro, T.G., Ed. Biological Applications of Raman Spectroscopy Volume 1; Wiley: New York; 1987.
39. Bertoluzza, M.; Fagnano, C.; Monti, P. Spectroscopy of Biological Molecules. State-of-the-Art. Proc. 3rd Eur. Conf., Esculapio: Bologna; 1989.
40. Painter, P.C. In Food Analysis: Principles and Techniques; Volume 2; Gruenwedel, D.W.; Whitaker, J.R., Eds.; Marcel Dekker: New York; 1984; Chapter 11.
41. Lord, R.C.; Yu, N.-T. J. Mol. Biol. 1970, 50, 509-524.
42. Chen, M.C.; Lord, R.C.; Mendelsohn, R. Biochim. Biophys. Acta 1973, 328, 252-260.
43. Chen, M.C.; Lord, R.C.; Mendelsohn, R. J. Amer. Chem. Soc. 1974, 96, 3038-3042.
44. Clark, A.H.; Lee-Tuffnell, C.D. In Functional Properties of Food Macromolecules; Mitchell; J.R.; Ledward, D.A., Eds.; Elsevier Applied Science: New York, 1986; Chapter 5.
45. Verma, S.P.; Wallach, D.F.H. Biochim. Biophys. Acta 1977, 486, 217-227.
46. Verma, S.P.; Wallach, D.F.H. Biochem. Biophys. Res. Comm. 1977, 74, 473-479.

RECEIVED August 16, 1990

Chapter 16

Excellent Emulsifying Properties of Protein–Dextran Conjugates

Akio Kato and Kunihiko Kobayashi

Department of Agricultural Chemistry, Yamaguchi University, Yamaguchi 753, Japan

Protein-dextran conjugates having excellent emulsifying properties have been developed by coupling proteins to CNBr-activated dextran or by linking proteins with dextran through naturally occurring Maillard reaction. The emulsifying properties of both protein-dextran conjugates were much higher than native proteins. Furthermore, these conjugates were superior to commercial emulsifiers from sucrose-fatty acid esters and polyglycerin esters, especially at high salt concentrations and at acidic pH. In addition, the emulsifying properties of protein-dextran conjugates were greatly enhanced by preheating at 100°C. Thus, it was suggested that protein-dextran conjugates may be useful as a macromolecular emulsifier for food or drug application where conditions of acidic or alkaline pH and heat-treatment are required.

Proteins have unique surface properties due to their large molecular weight and their amphiphilic properties. Proteins are generally unstable to heating for the pasteurization, to shaking and homogenization for the preparation of emulsion. Furthermore, some proteins coagulate during emulsifying process as a result of surface denaturation. Thus, many studies on the chemical and enzymatic modification of proteins have attempted to improve the emulsifying properties of proteins. We found that a soluble protein-dextran conjugate prepared by coupling proteins to cyanogen-bromide activated dextran exhibited emulsifying properties which were superior to commercial emulsifiers (1). This result suggested that a covalently linked protein-polysaccharide conjugate could be used to make new functional biopolymers. Since protein-polysaccharide conjugates formed by covalent attachment enhanced the stability and solubility of proteins, this technique for conjugate formation could be utilized for medical and food applications, if the safety of the conjugates were ensured. For these applications, the use of chemical reagents to make conjugates should be avoided. Here, we describe a

0097–6156/91/0448–0213$06.00/0
© 1991 American Chemical Society

safe protein-polysaccharide conjugate prepared without using chemical
reagents. One of the most promising way to conjugate is through
naturally occurring Maillard reaction. Since dextran has only one
active reducing-end per molecule, the formation of protein-dextran
conjugates is possible by the Maillard reaction. As expected,
protein-dextran conjugates were covalently attached by the Maillard
reaction through the protein amino groups and the reducing-end in
dextran. This work describes the properties of these protein-dextran
conjugates.

MATERIALS AND METHODS

Materials. Dextran (molecular weight, 60,000-90,000) was from
Wako Pure Chemical Industries, and Sephacryl S-300 was from Pharmacia
LKB. Ovalbumin was isolated and recrystallized five times from fresh
egg white with sodium sulfate (2). Lysozyme was prepared from fresh
egg white by a direct crystallization method (3) and recrystallized
five times. SunSoft SE-11 and Q-18S were supplied from Taiyo Kagaku
Co.(Japan).
    Preparation of ovalbumin-dextran conjugate by CNBr-activated
dextran. Ovalbumin-dextran conjugate was prepared by the method of
Marshall and Rabinowitz (4). To a stirred solution of dextran (2.5g)
in water (250ml), adjusted to pH 10.7 with 500mM sodium hydroxide
solution, cyanogen bromide (0.625g) was added, followed by a second
addition of cyanogen bromide (0.625g), 30 min later. The pH was
maintained at 10.7 during this process by addition of sodium
hydroxide solution (500mM). Thirty minutes after the second addition
of cyanogen bromide, the pH was adjusted to 9.0 by addition of 100mM
hydrochloric acid solution. After dialysis at 4 °C for 2 h against
4 liters of sodium carbonate solution, pH 9.0 (prepared by addition
of 1.0 M sodium carbonate solution to distillized water until the pH
reached 9.0), ovalbumin (0.50g) was added. The pH was maintained at
9.0 during addition of ovalbumin, by addition of sodium carbonate
solution (200mM). Coupling of ovalbumin to cyanogen bromide-activated
dextran was then allowed to proceed during 12 h at 4°C. After this
period, the solution was dialyzed for 2 h at room temperature against
4 liters of sodium carbonate solution (prepared as above), then 20 ml
of glycine solution (100mg/ml) was added. After standing for an
additional 12 h at 4°C, the product was lyophilized.
    Preparation of ovalbumin-dextran conjugate by Maillard reaction.
Freeze-dried ovalbumin-dextran mixtures in the weight ratio of 1:5
were stored at 60 °C and 65 % relative humidity for 3 weeks (5). To
further purify ovalbumin-dextran conjugate, gel filtration of the
conjugate was performed on a column (70 x 3 cm) of Sephacryl S-300.
Elution was carried out with 50 mM acetate buffer, pH 5.0, containing
10 mM sodium chloride, and 3.0 ml fractions were collected. The
conjugate peak (fraction number 26 to 36) was combined together,
dialyzed against deionized water and lyophilized.
    Determination of the molecular weight of ovalbumin-dextran
conjugate. 0.1% ovalbumin-dextran conjugate solution in 100 mM sodium
phosphate buffer (pH 7.0) containing 0.1 % SDS was applied to a high
performance gel chromatography system, employing a TSK gel G3000SW
gel column(Toyo Soda Co., 0.75 x 60 cm) at a flow rate of 0.3 ml/min.
Effluent from the column was monitored with a low-angle laser light
scattering photometer (LS-8,Toyo Soda Co.) and then with a precision

differential refractometer(RI-8,Toyo Soda Co.). The molecular weight
of ovalbumin-dextran conjugate was estimated from the ratio of total
area in the peak of a low-angle laser light scattering photometer(LS)
to that of a refractometer (RI) by the method of Takagi and Hizukuri
(6).
   Measurement of emulsifying properties. Emulsifying properties
were measured by the method of Pearce and Kinsella(7). To prepare
emulsions, 1.0 ml of corn oil and 3 ml of protein solution in 0.1 M
phosphate buffer, pH 7.4, were shaken together and homogenized in an
Ultra Turrax (Hansen & Co., West Germany) at 12,000 rpm for 1min at
20 °C. 50 μl of emulsion was taken from the bottom of the container
after different times and diluted with 5ml of 0.1 % sodium dodecyl
sulfate solution. The absorbance of diluted emulsion was then
determined at 500nm. The emulsifying activity was determined from
the absorbance measured immediately after emulsion formation. The
emulsion stability was estimated by measuring the half-time of the
turbidity of emulsion.
   SDS-slab polyacrylamide gel electrophoresis. SDS-slab polyacrylamide gel electrophoresis was carried out by the method of
Laemmli(8) using 10 % acrylamide separating gel, and 3 % stacking
gel containing 0.1 % SDS. Protein samples (20μl, 0.1 %), were
prepared in Tris-glycine buffer, pH 8.8, containing 1 % SDS and 1 %
mercaptoethanol. Electrophoresis were carried out at constant current
of 10 mA for 5 h using electrophoretic buffer of Tris-glycine
containing 0.1 % SDS. The gel sheets were stained for proteins and
carbohydrates with Coomassie blue G-250 and Fuchsin, respectively.
   Measurement of lysozyme activity. Both lysis and hydrolysis
were measured using Micrococcus lysodeikticus and glycol chitin as
substrates, respectively. Lytic activity was estimated from the
decrease in the turbidity of Micrococcus lysodeikticus cell
suspension and the hydrolysis activity was calculated from the
increase in the reducing group of hydrolyzed N-acetylglucosamine
(9).
   Determination of amino groups in protein-dextran conjugate. The
content of free amino groups in the conjugate was determined by the
method of Haynes et al.(10) using a specific reagent for amino
groups, trinitrobenzene sulfonate.

RESULT & DISCUSSION

   A scheme for the preparation and the binding mode of protein-
dextran conjugates is shown in Fig.1. When protein is coupled with
CNBr-activated dextran, polymerized networks are formed between
functional amino groups in protein and numerous activated hydroxyl
groups contained in dextran molecule. On the other hand, when
powdered protein-dextran mixture is stored at 60 °C and at 65-79 %
relative humidity, conditions optimal for the Maillard reaction, only
one or two moles of dextran link per moles of protein. Because there
is only one reducing-end group per dextran molecule, dextran and
protein react without the formation of a network structure. Evidence
for this hypothesis was obtained from measuring the molecular weight
of ovalbumin-dextran conjugates. The molecular weight of conjugates
prepared from CNBr-activated dextran were 250,000-2,000,000, whereas
those prepared from the Maillard reaction were 130,000 - 230,000
(Table I).

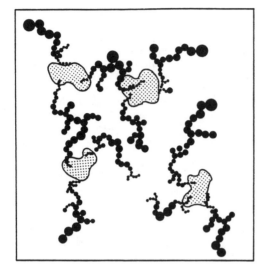

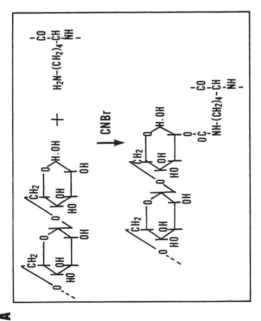

Figure 1. Scheme for the binding reaction (left) and the binding mode (right) of ovalbumin-dextran conjugates through CNBr-activated dextran (A) and Maillard reaction (B). Dotted areas designate protein molecules whereas the branched solid circles designate dextran molecule.

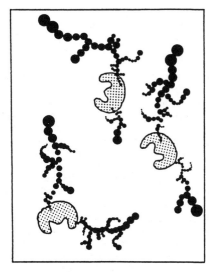

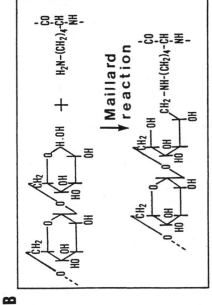

Figure 1. Continued.

Table I.   Molecular Weight Distribution of Ovalbumin-Dextran
           Conjugates Estimated by Low-Angle Laser Light
           Scattering Technique

Ovalbumin-dextran conjugates	Molecular weight distribution
CNBr-activated dextran	250,000 - 2,000,000
Maillard reaction	130,000 - 230,000

SDS-polyacrylamide gel electrophoretic patterns led to the same
conclusion (Fig.2). The electrophoretic pattern of the ovalbumin-
dextran conjugate prepared by the Maillard reaction shows a single
band for protein and carbohydrate stains near the boundary between
stacking and separating gels, whereas the ovalbumin-dextran conjugate
prepared from CNBr-activated dextran showed a broad spectrum of high
molecular weight bands in the stacking gel,indicating a polydispersed
high molecular distribution.
     As shown in Fig.3-5, the emulsifying properties of protein-
dextran conjugate prepared from CNBr-activated dextran were compared
with those of commercial emulsifiers. Emulsion turbidity(ordinate) is
plotted against the standing time after emulsion formation(abscissa).
The commercial emulsifiers were SunSoft SE-11 and Q-18S. Sunsoft SE-
11 is a sucrose-fatty acid ester whereas Q-18S is a polyglycerin
ester. Emulsifying properties were compared in two emulsion systems
both of which had equal volumes of water and oil (Fig.3). In one
system the emulsifier was added to the water phase whereas in the
other system it was added to the oil phase. The ovalbumin-dextran
conjugate was comparable to the commercial emulsifiers and was found
suitable for both emulsion system. Apparently, ovalbumin-dextran
conjugates can be used for either oil in water or water in oil
emulsions. In Fig.4 is shown the effect of salt on the emulsifying
property of the oil in water emulsion. 10 % NaCl was contained in the
water phase. The relative emulsifying activity, which was determined
from the emulsion turbidity immediately after emulsion formation, of
ovalbumin-dextran conjugate, SE-11 and Q-18S were 0.734, 0.130 and
0.077, respectively. Comparison of the commercial emulsifiers with
the ovalbumin-dextran conjugate revealed that the conjugate had
better emulsifying properties in the presence of 10 % NaCl. In Fig.5
is shown the effect of acid on the emulsifying property of the oil in
water emulsion. The pH of the water phase was lowered to 2.3 with 1 %
citric acid. In the presence of acidic salt, the ovalbumin-dextran
conjugate was stable. The value of relative emulsifying activity of
ovalbumin-dextran conjugate, SE-11 and Q-18S were 0.630, 0.150 and
0.235, respectively. Stable emulsifying properties in the acidic pH
region are important for industrial application, in that the
emulsifying properties of most commercial emulsifiers are greatly
reduced in the low pH region.
     In Fig 6-8, the emulsifying properties of ovalbumin-dextran
conjugate prepared by both the Maillard reaction and CNBr-activated
dextran are compared.   In Fig.6 are compared the emulsifying
properties of various ovalbumin-dextran conjugates in 0.1 M phosphate
buffer, pH 7.4. The emulsifying activity(the turbidity at 0 time) and
emulsion stability (half-life of emulsion turbidity) of ovalbumin-

PROTEIN STAIN

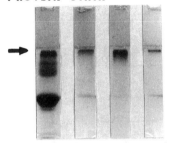

CARBOHYDRATE STAIN

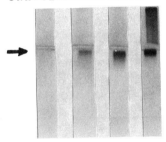

1   2   3   4

Figure 2. SDS-Polyacrylamide gel electrophoretic patterns of
ovalbumin-dextran conjugates. lane 1, ovalbumin stored at 60 °C
for 3 weeks; lane 2,ovalbumin-dextran conjugate obtained by dry-
heating at 60°C for 3 weeks; lane 3, ovalbumin-dextran conjugate
purified from lane 2; lane 4,ovalbumin-dextran conjugate prepared
by CNBr-activated dextran.
(Reproduced with permission from Ref. 5. Copyright
1990 Japan Society for Bioscience, Biotechnology,
and Agrochemistry.)

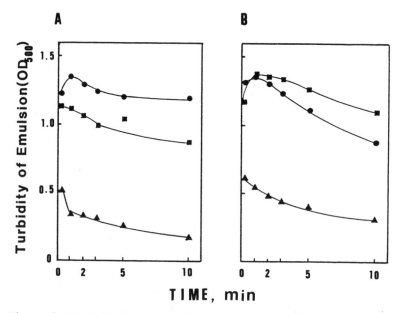

Figure 3. Emulsifying properties of ovalbumin-dextran conjugate
and commercial emulsifier in the emulsion systems where
emulsifier were added to water phase (A) and oil phase (B).
● , 0.1% ovalbumin-dextran conjugate; ■ , 0.1% Sunsoft SE-11;
▲ , 0.1% Sunsoft Q-18S.
(Reproduced from Ref. 1. Copyright 1988 American
Chemical Society.)

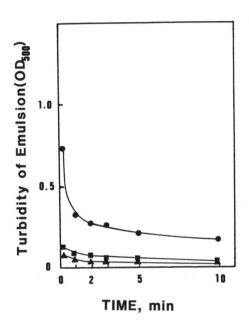

TIME, min

Figure 4. Effect of salt on the emulsifying properties of ovalbumin-dextran conjugate and commercial emulsifiers in oil in water emulsion(O/W = 1/3), with 10 % NaCl contained in the water phase. ●, 0.1% ovalbumin-dextran conjugate;■,0.1% Sunsoft SE-11; ▲ , 0.1% Sunsoft Q-18S.
(Reproduced from Ref. 1. Copyright 1988 American Chemical Society.)

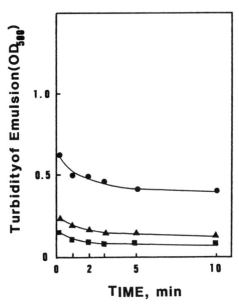

Figure 5. Effect of acid on the emulsifying properties of
ovalbumin-dextran conjugate and commercial emulsifiers in oil in
water emulsion (O/W = 1/3) with 1 % citric acid contained in
the water phase, pH 2.3. ●, 0.1% ovalbumin-dextran conjugate;
■, 0.1% Sunsoft SE-11; ▲, 0.1% Sunsoft Q-18S.
(Reproduced from Ref. 1. Copyright 1988 American
Chemical Society.)

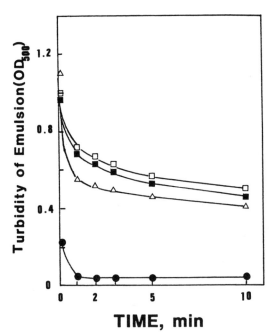

Figure 6. Emulsifying properties of ovalbumin-dextran conjugate
obtained by dry-heating at 60 °C for 3 weeks. □, faction eluted
in void volume on a gel filtration with Sephacryl S-300; ■ ,
fraction eluted later void volume on a Sephacryl S-300; △ ,
ovalbumin-dextran conjugate prepared by CNBr-activated dextran.
● , mixture of ovalbumin with dextran (control).
(Reproduced with permission from Ref. 5. Copyright
1990 Japan Society for Bioscience, Biotechnology,
and Agrochemistry.)

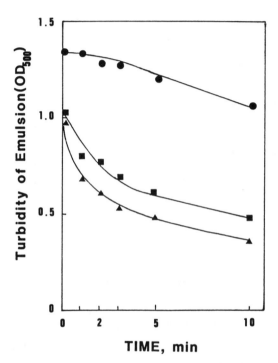

TIME, min

Figure 7. Effects of various pH on the emulsifying properties of
ovalbumin-dextran conjugate obtained by dry-heating at 60 °C for
3 weeks. ■ , pH 7.4 (1/15 M phosphate buffer); ● , pH 10 (1/15 M
carbonate buffer); ▲ , pH 3 (1/15 M citrate buffer).
(Reproduced with permission from Ref. 5. Copyright
1990 Japan Society for Bioscience, Biotechnology,
and Agrochemistry.)

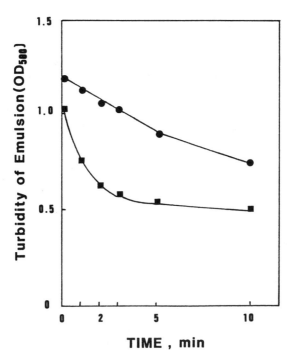

TIME , min

Figure 8. Effects of heating sample on the emulsifying properties
of ovalbumin-dextran conjugate obtained by dry-heating at 60 °C
for 3 weeks. ■ , unheated ovalbumin-dextran conjugate; ● ,
ovalbumin-dextran conjugate heated at 100 °C.
(Reproduced with permission from Ref. 5. Copyright
1990 Japan Society for Bioscience, Biotechnology,
and Agrochemistry.)

dextran conjugates were higher than those of native ovalbumin. The
emulsion stability of the ovalbumin-dextran conjugate obtained by
Maillard reaction was higher than that of the conjugate prepared from
CNBr-activated dextran. A slight difference was observed between
crude and purified conjugates prepared by the Maillard reaction. In
Fig.7 is shown the effect of various pH on the emulsifying properties
of the ovalbumin-dextran conjugate prepared by the Maillard reaction.
The high emulsifying properties of ovalbumin-dextran conjugate were
maintained as low as pH 3, and were further improved at pH 10. In
Fig. 8 is shown the effect of heating the sample on the emulsifying
properties of the ovalbumin-dextran conjugate prepared by Maillard
reaction. The emulsifying properties of ovalbumin-dextran conjugate
were greatly increased by preheating the conjugate at 100 °C in that
no insoluble matter was observed. Thus, another advantage of the
ovalbumin-dextran conjugate is its heat resistance which enables it
to be pasteurized. We have shown that ovalbumin-dextran conjugates
prepared by the Maillard reaction and by CNBr-activated dextran had
excellent emulsifying properties. Because of the preparation without
chemical reagents, the conjugate formed by the Maillard reaction is
more suitable for industrial applications than the conjugate formed
from CNBr-activated dextran. The conjugate prepared without chemicals
can be used as an emulsifier and as a protein food additive with heat
stability.
    Another protein-dextran conjugate with excellent functional
properties is that of lysozyme and dextran which were naturally
linked through the Maillard reaction. The data in Fig.9 revealed the
progressive lysozyme-dextran conjugation during the course of a
Maillard reaction over a period of 3 weeks. As shown in Fig.10,
emulsifying properties of lysozyme-dextran are greatly increased as
Maillard reaction proceeds. Lysozyme activity was considerably
restored in lysozyme-dextran conjugate (Table II) and the heat
stability was greatly enhanced by conjugate formation (Table III).

Table II. Enzymatic Activity of Native Lysozyme and
           Lysozyme-Dextran Conjugate Prepared by
           Maillard Reaction

Lysozyme	Substrate	
	Glycol chitin	M. lysodeikticus
Native	100 %	100 %
Conjugate	80 %	32 %

Table III. Heat Stability of Lysozyme and Lysozyme-Dextran
            Conjugate Prepared by Maillard Reaction

Lysozyme	Half life (min)
Native	15
Conjugate	40

Half life is represented as the time when
lytic activity lowered to 50 % on heating
at 100 °C.

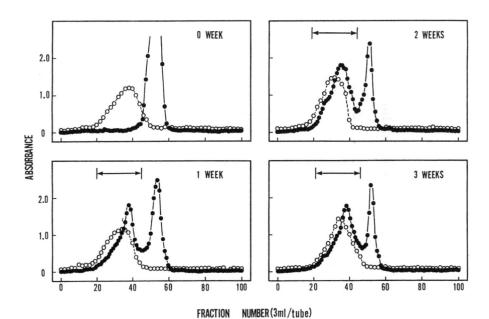

FRACTION NUMBER (3ml/tube)

Figure 9. Elution patterns in 50 mM acetate buffer (pH 5) on a Sephacryl S-300 column of lysozyme-dextran mixtures incubated at 60 °C for 0-3 weeks under 78 % relative humidity. ●——● , absorbance at 280 nm (for protein); o——o, absorbance at 470 nm to follow the color development by phenol-sulfate method (for carbohydrate).

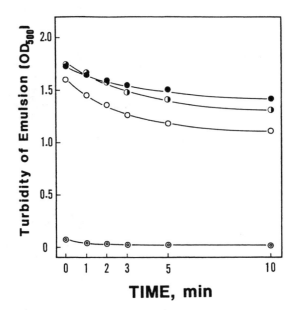

Figure 10. Emulsifying properties of lysozyme-dextran conjugate purified by gel filtration on Sephacryl S-300. ◉ ,   0 week; ○ , incubated at 60 °C for 1 week; ● , incubated at 60°C for 2 weeks; ◐ , incubated at 60 °C for 3 weeks.

The binding mode was studied in detail. Two free amino groups were found to be reduced in lysozyme-dextran conjugate(Table IV). This suggests that two moles of dextran attach to one mole of lysozyme. This binding mode is supported from the data in the binding ratio of dextran to lysozyme, as shown in Table V.

Table IV. Contents of Free Amino Group in Lysozyme-Dextran Conjugate Prepared by Maillard Reaction

Lysozyme	Numbers of free amino groups per mole
Native	7.0
Conjugate	4.8

Table V. Binding Ratio of Dextran to Lysozyme in Conjugate Prepared by Maillard Reaction

	Lysozyme	Dextran
Weight ratio	1	11.3
Molar ratio	1	2.1

Thus limited binding of dextran to lysozyme may result from the steric hindrance because of macromolecular interaction between them. This is also the case of ovalbumin-dextran conjugate (5). Lysozyme-dextran conjugate without the use of chemicals can be potentially used for food additive possessing bifunctional properties, either emulsifier or antimicrobial reagent.

As mentioned above, the protein-dextran conjugates can be useful for the industrial application as a new functional emulsifier that is soluble, macromolecular, heat-stable, and stable in acid or in high salt concentration. Finally, it may be possible to make other functional protein food additives by selecting appropriate combinations of proteins with polysaccharides using the reactions described here.

LITERATURE CITED

1. Kato,A.; Murata,K.; Kobayashi,K. J. Agric. Food Chem. 1988,36, 421-425.
2. Kekwick,R.A.; Cannan,R.K. Biochem. J. 1936, 30, 277-280.
3. Alderton,G.; Fevold,H.L. J. Biol. Chem. 1946, 164, 1-5.
4. Marshall,J.J.;Rabinowitz,M.L. J. Biol. Chem. 1976, 251, 1081-1087.
5. Kato,A.; Sasaki,Y.; Furuta,R.; Kobayashi,K. Agric. Biol. Chem. 1990, 54, 107-112.
6. Takagi,T. Hizukuri,S. J. Biochem. 1984, 95, 1459-1467.
7. Pearce,K.N.;Kinsella,J.E. J. Agric. Food Chem. 1978, 26, 716-723.
8. Laemmli,U.K. Nature 1970, 227, 680-685.
9. Imoto,T.; Yagishita,K. Agric. Biol. Chem. 1971, 35, 1154-1156.
10. Haynes,R.; Osuga,D.T.; Feeney,R.E. Biochemistry 1967, 6, 541-548.

RECEIVED August 16, 1990

Chapter 17

Effect of Polysaccharide on Flocculation and Creaming in Oil-in-Water Emulsions

Margaret M. Robins

AFRC Institute of Food Research, Colney Lane, Norwich NR4 7UA,
United Kingdom

Creaming of the droplets in 20% alkane-in-water emulsions
in the presence and absence of non-adsorbing polymer is
reported in the form of oil concentration-height profiles
collected at intervals of time. Without polymer, the
polydisperse droplets (stabilised by non-ionic surfactant) cream individually at a rate determined by their
diameter. Measurement of creaming rates enables the
distribution of hydrodynamic diameters to be inferred, and
it agrees well with the size distribution from a light
diffraction method. In the presence of hydroxyethylcellulose in the continuous phase at a concentration exceeding 0.03%w/w, the droplets flocculate, and the creaming rates are used to estimate the size of the pores in
the flocculated droplet network. At 0.03% polymer, flocculated and individual particle phases are in coexistence. Direct evidence for a depletion mechanism of
flocculation is presented.

Many foods are emulsions during or after manufacture (1). In oil-
in-water emulsions, the dispersed oil droplets generally possess a
lower density than the continuous aqueous phase. Unless the
droplets are very small or very concentrated the density difference
leads to the accumulation of the droplets at the top of the
container ("creaming") with consequent loss of perceived quality.
This paper presents results on the creaming of oil-in-water
emulsions in the absence and presence of the polysaccharide
hydroxyethylcellulose (Natrosol 250HR).

Methods

Emulsion preparation and characterisation. Emulsions were prepared
containing droplets of mixed alkanes (heptane and hexadecane in the
volume ratio 90:10), stabilised to coalescence by the non-ionic
surfactant Brij 35 at a concentration of 0.35%w/w in the final
continuous phase. The emulsions were initially prepared in a Waring

0097–6156/91/0448–0230$06.00/0
© 1991 American Chemical Society

Blendor at 60% oil volume fraction ϕ, and diluted to obtain ϕ = 20%. The diluent was an aqueous solution of preservative alone (sodium metabisulphite at 0.2%w/w in the final continuous phase), or preservative and polymer (Natrosol 250HR). The droplet size distribution was measured using a Malvern Mastersizer light diffraction instrument. Since the same concentrated emulsion was used the droplet size distribution was identical in all the samples.

<u>Characterisation of disperse and continuous phases.</u> The density of the liquids was measured at 20°C using an Anton Paar DMA602 vibrating tube density meter.

Viscosities were measured at low shear-rate at 20°C using a double-gap measuring system fitted to a Bohlin Reologi Controlled Stress rheometer.

The concentration of polymer in the "sub-cream" layer of continuous phase formed after each emulsion had creamed was determined using a spectrophotometric method (<u>2</u>). The sub-cream layers were diluted to a nominal concentration of 0.006% v/v and 0.5ml pipetted into a stoppered test-tube, to which was added 1.5ml 3% w/w phenol solution and 5ml concentrated sulphuric acid. The mixture was shaken and left to stand for 20 minutes, before samples were transferred to cuvettes for measurement of absorbance at 488nm. Standard solutions of 0.005 and 0.007% w/w polymer were also measured, for calibration purposes. The error on polymer concentration was less than 10%.

<u>Measurement of creaming in emulsions.</u> Most techniques to monitor creaming (or sedimentation) are intrusive, such as sampling (<u>3</u>), or applicable only to dilute systems (<u>4</u>). Recently several non-intrusive methods have been developed (<u>5</u>), including the use of ultrasonics. We have developed a technique based on the velocity of ultrasound through the dispersion, which may be directly related to its composition (<u>6</u>). The technique is suitable for non-aerated dispersions with a wide range of concentrations (>1%v/v), it is non-destructive and non-intrusive. The apparatus, shown schematically in Figure 1, determines the velocity of ultrasound from measurements of the time-of-flight of a pulse of ultrasound generated from a continuous wave of frequency 6.4MHz. The time is determined to a precision of 5ns in a typical propagation time of 25μs. The transducers are held at fixed separation and moved vertically so that measurements are made at a series of heights with a spatial resolution of less than 2mm. The samples are contained in parallel-sided cells (dimensions typically 16mm wide x 32mm deep x 160mm high) of polymethylmethacrylate. The cells and transducers are immersed in a water bath held at 20°C to maintain a constant temperature and to provide a good ultrasonic contact between the transducers and the cell.

In general the velocity of ultrasound through a dispersion at a given frequency is a complex function of the composition, particle size distribution and the physical properties of the dispersed and continuous phases (<u>7</u>). However, in many simple dispersions the particle size and state of aggregation of the particles have a negligible effect on the ultrasonic velocity, and the overriding factor is the particle concentration. Often a simple mixing equation (<u>8</u>) describes the relationship between the ultrasonic velocity V and volume fraction ϕ

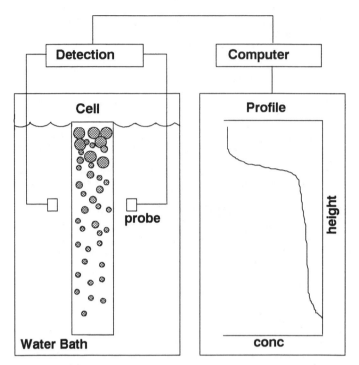

Figure 1:
Schematic diagram of ultrasonic creaming monitor.

$$V^2 = \frac{V_c{}^2}{\left[1-\phi\left[1-\frac{\rho_d}{\rho_c}\right]\right]\left[1-\phi\left[1-\left[\frac{V_c}{V_d}\right]^2\frac{\rho_c}{\rho_d}\right]\right]} \qquad (1)$$

where V is the velocity of ultrasound through the dispersion of volume fraction ϕ, and ρ_c, ρ_d, V_c, V_d are the densities of and ultrasound velocities through the continuous and disperse phases. Figure 2 shows the measured and predicted velocity through alkane-in-water emulsions with a range of oil concentration. The simple mixing equation (1) is clearly a good model for this system.

Theory of Creaming or Sedimentation

The terminal velocity v_s of a single spherical particle moving under gravity in a viscous liquid is given by Stokes' Law

$$v_s = \frac{\Delta\rho.d^2.g}{18\ \eta_c} \qquad (2)$$

Velocity (v/ms⁻¹)

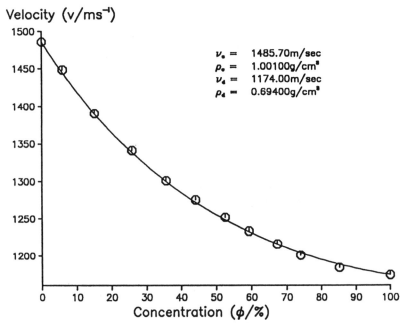

Figure 2:
Ultrasonic velocity and oil concentration.   Line from equation
(1).

where $\Delta\rho=\rho_c-\rho_d$, the density difference between the continuous and
disperse phases, d is the droplet diameter, g, the acceleration due
to gravity and η_c, the viscosity of the continuous phase.   When
applying the simple analysis to non-dilute emulsions we have to make
allowance for the effect of the other droplets on the creaming rates
of individual droplets.   This influence may be incorporated in the
theory in two ways.
     It is well known that the presence of particles in a dispersion
increases its viscosity.   One approach is to quantify the increase
in η_c and apply equation (2) to obtain the decrease in v_s.   A mean-
field model of dispersion viscosity (9) proposes a relationship
between the viscosity η of a dispersion of volume fraction ϕ and the
continuous phase viscosity η_0

$$\eta = \eta_0 \ (1-\phi/\phi_m \ )^{-2.5 \ \phi_m} \tag{3}$$

where ϕ_m is the maximum (close-packed) volume fraction of the particles.   For an emulsion of 20% volume fraction with ϕ_m = 70%,
equations (2) and (3) predict v = 0.55v_s.
     There are also empirical correlations for the effect of concen-
tration on the velocity of a monodisperse suspension, where the
particles move as a body.   That of Richardson and Zaki (10) predicts
v = 0.35v_s.   It is debatable which expression is valid for
polydisperse droplets moving at different speeds through an
emulsion; we use equation (3) in preference to an empirical
correlation.

Creaming of Emulsions Without Polymer

    In Figure 1 a typical oil concentration profile is shown
schematically alongside the creaming emulsion. The majority of the
oil is at the top of the cell in a concentrated cream layer, with
the slower-moving droplets beginning to clear from the base. This
is typical behaviour for a polydisperse emulsion containing no
polymer. Figure 3 shows (in a horizontal format) concentration
profiles at various stages during the creaming of a polymer-free
emulsion. Creaming is detected within 2 hours, although the
emulsion appears uniform visually for over 10 days. Initially the
oil is uniformly dispersed in the container, at a concentration ϕ_0 =
20% v/v. After a few days the droplets have started to rise up the
cell, so the concentration at the base has fallen, and there is a
concentrated cream layer at the top. With time, all the droplets
arrive at the top.   If they were monodisperse, the lower meniscus
would be sharp, as they would all cream at the same speed. It is
possible to use the observation that they move at different speeds
(resulting in the hazy, diffuse "meniscus" rising up the cell), to
obtain the effective hydrodynamic size distribution (11).

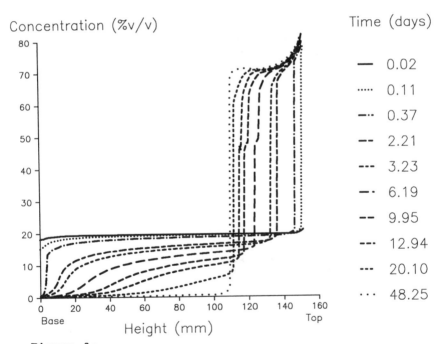

Figure 3:

Oil concentration profiles for 19% alkane emulsion without
polymer.

<u>Determination of droplet size distribution</u>. We assume that the
droplets travel individually to the cream layer at a speed
consistent with their diameter and Stokes' Law (equation (2), with
the mean-field correction for viscosity, equation (3)). It is
possible to identify particle fractions of different size by the
changing shape of the lower, diffuse meniscus. Consider the
position on the meniscus at which the oil concentration is 10%,
approximately half the total oil concentration. The rate at which
this 10% contour rises up the cell is related by Stokes' Law to a
particular droplet diameter, d_m. Since half the oil has already
moved away from the meniscus, d_m represents the mid-point diameter,
the median diameter. Similarly, the rate of rise of the contour at
5% oil represents the lower quartile droplet diameter. The heights
of typical contours are shown as a function of time in Figure 4.
Each is linear, showing a constant velocity of each fraction of
droplets of diameter d. The velocity of each contour enables a
cumulative size distribution to be inferred, as shown in Figure 5.

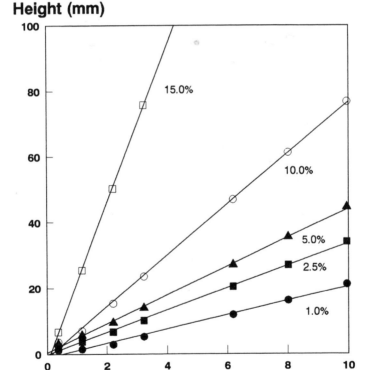

Figure 4:
Height of oil concentration contours and creaming time for
emulsion without polymer.

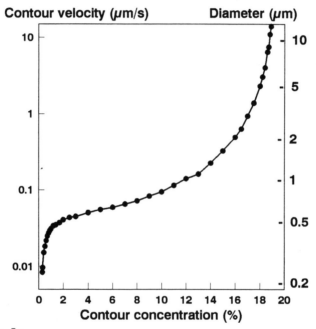

Figure 5:
Velocity of contours and inferred hydrodynamic diameters for each
concentration contour.

The resulting size distribution is shown in Figure 6, and compared
with that obtained from a laser diffraction technique, the Malvern
Mastersizer. The agreement is very good, considering that the
creaming method is subject to several assumptions as to the
independence of the motion of the particles, and the problem of
allowing correctly for the local droplet concentrations.

Creaming of Emulsions Containing Polymer

Polysaccharide stabilisers are frequently used to reduce the rate of
creaming, as well as to impart the required mouthfeel properties to
the product. We are interested in the mechanisms by which the
polymers influence the separation process. In particular, previous
work (12-13) has shown that if insufficient polymer is added, the
droplets become flocculated and cream faster than in the absence of
the stabiliser. Here we present creaming results on 20% alkane-in-
water emulsions containing the polysaccharide hydroxyethylcellulose
(Natrosol 250HR). The droplet size distribution is the same as in
the polymer-free emulsion (Figure 6).
      At concentrations up to and including 0.02%w/w in the
continuous phase the creaming is not visible to the eye until nearly
complete, and the creaming profiles are very similar to those
without polymer (Figure 3). However, at a level of 0.03%, there is
a change in the creaming behaviour. The profiles are shown in
Figure 7. Visually the emulsion appears opaque until after about
10 days, when the base starts to clear of oil. However the ultrasonic profiles show that there are two distinct types of creaming.
The majority of the oil droplets move up the cell rapidly, with a
sharp meniscus, but about 4% oil remains to cream slowly with a
diffuse meniscus, as in the emulsions with zero or low levels of
polymer.
      The contour heights for the emulsion containing 0.03% polymer
are shown in Figure 8. At contour concentrations below 4%, the
droplets are moving in fractions, as they did when no polymer was
present. The range of contour velocities exhibited indicates that
the droplets moving individually represent a similar range of
diameters as the original emulsion. However, they remain as
individual droplets while the majority of the oil creams very fast,
with little variation in speed with contour concentration. This
fraction is clearly aggregated, and the droplets are creaming as a
single entity, presumably in a network with sufficiently large voids
for the unaggregated fraction to drain through. We have thus
observed coexistence of two emulsion phases; a flocculated fraction
containing 80% of the oil, and an unflocculated fraction which can
cream as individual droplets.
      When the polymer concentration is increased further, several
effects are apparent. All the oil appears to become aggregated,
creaming rapidly with a sharp meniscus. Although the viscosity of
the continuous phase increases with polymer concentration, the
creaming is faster than without polymer. Figure 9 shows the
concentration profiles for an emulsion containing 1% polymer in the
continuous phase. The viscosity (at low shear-rate) of the polymer
solution is 3.5Pas, but the creaming rate of the meniscus is
$0.18\mu m.s^{-1}$, faster than the 10% contour in the polymer-free system.

Porous network model. The sharpness of the meniscus and the shape
of the concentration profiles above the meniscus are reminiscent of
the behaviour of a solid material under compression. We thus model

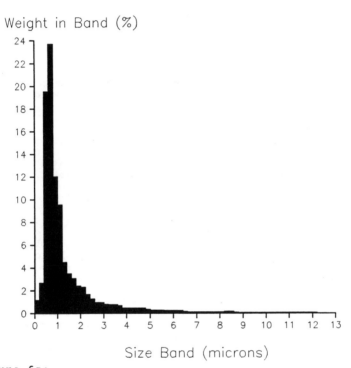

Figure 6a:
Droplet size distribution inferred from concentration
contours.

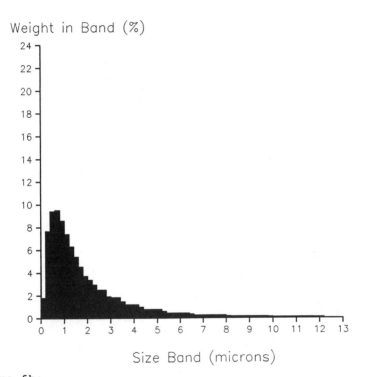

Figure 6b:
Droplet size distribution from light diffraction particle
sizer (Malvern Mastersizer).

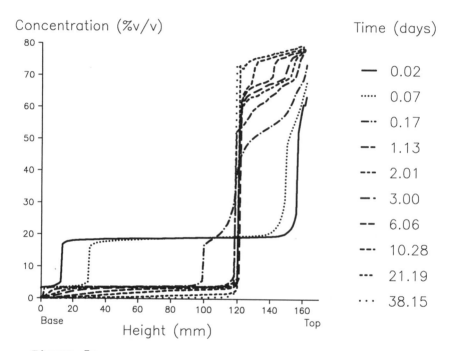

Figure 7:
Oil concentration profiles for 20% alkane emulsion containing
0.03%w/w polymer in the continuous phase.

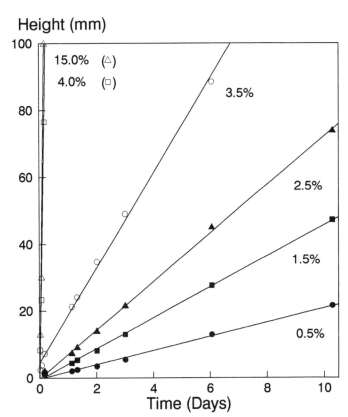

Figure 8:
Height of oil concentration contours and creaming time for
emulsion containing 0.03%w/w polymer in the continuous phase.

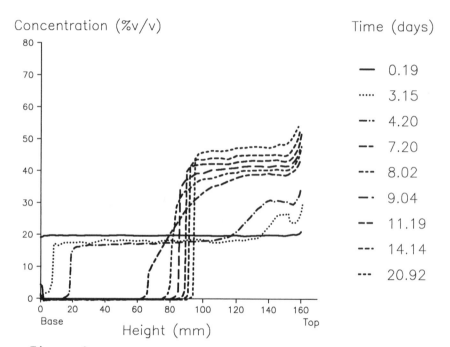

Figure 9:
Oil concentration profiles for 20% alkane emulsion   containing
1.0%w/w polymer in the continuous phase.

the flocculated emulsion as an open but continuous network, which compresses under gravity. The compression results in the continuous phase draining through channels in the network. The creaming/compression rate is limited by the viscous resistance of the continuous phase as it flows through the channels. The drainage of liquid through a porous bed of particles has been studied extensively, and the applicability of the model to flocculated dispersions has been demonstrated by Michaels and Bolger (14), who investigated the sedimentation behaviour of flocculated kaolin suspensions. Their system differs in two main respects from ours; their particles are rigid, and they are very strongly flocculated. Emulsion droplets are deformable, and, in our experience, the flocs formed are weak and easily disrupted upon dilution. However, we have found the general descriptions of flocculated systems to be a useful starting point for analysis of our results.

The system is modelled as a porous bed through which the background liquid drains. The maximum sedimentation (or creaming) rate v gives an estimate of the effective hydrodynamic diameter of the drainage channels (pores), assumed to be smooth straight cylinders

$$v = \frac{\phi_0 . \Delta\rho . d_p^2 . g \ (1-\phi_f)}{32 \ \eta_c} \tag{4}$$

where ϕ_f = volume fraction of flocs $(1 - \phi_0/\phi_m)$ and d_p is the effective hydrodynamic pore diameter. This diameter was observed by Michaels and Bolger (14) to increase with strength of flocculation.

Applying the pore/channel model, we can estimate the diameter of the pores d_p in equation (4) from the creaming rate, v, viscosity η_c and oil concentrations ϕ_0, ϕ_m. We assumed that the droplets within the flocs were "close-packed", with $\phi_m = 0.7$. Figure 10 shows the estimated pore diameter as a function of polymer concentration. At low concentrations, the effective diameter of the pores is constant, at about 37μm. This is consistent with the observation in the coexistent emulsion (0.03% polymer) that individual droplets of up to 10μm could pass through the flocculated network. At higher levels, there is a large increase with concentration up to about 300μm in 1% polymer. This is consistent with stronger flocculation at higher polymer concentrations (14).

The increased strength of the flocs with higher levels of polymer is also evident in the reluctance of the cream layer to become closely packed. The oil concentration at the top of the samples in the early stages of creaming is indicative of the resistance of the flocs to compaction under gravity. With less than 0.03% polymer the cream is at 70% concentration from the start, but in the flocculated systems it builds up at much lower packing density. At the highest polymer concentration, (Figure 9) the cream is almost uniform at $\phi \sim 40$% before exhibiting slow compaction with time, again uniformly.

Depletion Flocculation. We have observed that at a critical concentration of polymer, 0.03%w/w, a flocculated phase and individual droplets can coexist. This is consistent with a depletion rather than bridging mechanism (15-16). Unless a polymer is attracted to the surface of the droplets, and thus becomes adsorbed, geometrical constraints near the droplet require that the density of polymer segments near the surface is lower than in the bulk continuous phase. The region near the droplets is thus depleted of polymer. If one visualises depletion flocculation as being driven by a need to reduce the volume of continuous phase from which the polymer is excluded, the occurrence of flocculation at the critical concentration should depend on the volume fraction of droplets. This image is supported by an experiment using 0.03% polymer but with only 5% oil volume fraction. No flocculated fraction was observed, all the droplets creaming individually as in the systems with less polymer. The critical concentration, around 0.03%, is consistent with the experiments and calculations of Sperry (15-16), for the same polymer and the same mean particle size.

Another consequence of depletion flocculation is that the final continuous phase, contained in the sub-cream layer, should contain slightly more polymer than the overall concentration, because the continuous phase in the cream layer with depleted. For this reason we analysed the sub-cream layers for polymer using density and spectrophotometry. Figure 11 shows the estimated polymer concentra-

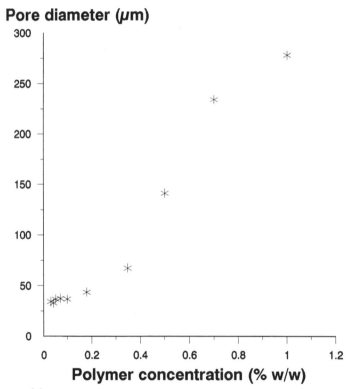

Figure 10:
Estimated diameter of pores in flocculated emulsions and polymer concentration.

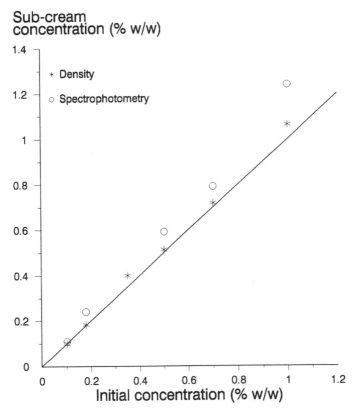

Figure 11:
Concentration of polymer in sub-cream layers and original polymer concentration. Line represents y=x.

tion in the sub-cream layers. (The density results, which depend on surfactant concentration in addition to polymer concentration, have been corrected to allow for the surfactant present on the droplets). There is clearly an enrichment over the initial, overall continuous phase concentration of polymer, although the precise amount is debatable. Taken as a whole, the evidence is strongly in favour of depletion flocculation. The weakness of the flocs, and the improbability that the polymer could adsorb onto the surfactant-coated droplets, are consistent with the depletion model.

Conclusions.

There is a considerable amount of information to be gained from measurement of concentration profiles during the creaming of emulsions. The rate of rise of the meniscus in flocculated systems

yields information on the strength of the flocculation, and we have
fitted a simple network model to our data. The size of drainage
pores in the network is inferred from the creaming/compression
rates. The critical polymer concentration required to flocculate
the droplets, the coexistent phases observed at that concentration
and the absence of flocculation at lower oil concentration ϕ_0, all
indicate that flocculation is by a depletion mechanism rather than
by bridging. The apparent enrichment by polymer of the sub-cream
layers provides direct evidence for its partial expulsion from the
interstices between the close-packed droplets.

Acknowledgments

The author is very grateful to Dr Andrew Howe, now at Kodak Research
Laboratories, Harrow, for his involvement in the early stages of the
work, and for helpful discussions since his departure from the
Institute. Technical assistance from David Hibberd, Paul Gunning,
Sarah Gouldby and Annette Fillery-Travis, and financial support from
the Ministry of Agriculture, Fisheries and Food are also gratefully
acknowledged.

Literature Cited

1. Dickinson, E.; Stainsby, G. Colloids in Food; Applied Science,
    Essex, 1982.
2. Dubois, M.; Gilles, K.A.; Hamilton, J.A.; Rebers, P.A.; Smith,
    F. Anal. Chem., 1956, 28, 30.
3. Reddy S. R.; Fogler, H. S. J. Coll. Int. Sci. 1981,
    82, 128.
4. Dickinson, E.; Murray, B. S.; Stainsby, G. Lebensm.
    Wiss. Technol., 1989, 22, 25.
5. Williams, R. A.; Xie, C. G.; Bragg, R.; Amarasinghe,
    W. P. K. Colloids Surfaces, 1990, 43, 1.
6. Howe, A. M.; Mackie, A. R.; Robins, M. M. J. Disp.
    Sci. Tech., 1986, 7, 231.
7. Allegra, J.R.; Hawley, S.A. J. Acoust. Soc. Am., 1972,
    51, 1545.
8. Urick, R. J. J. Appl. Phys., 1947, 18, 983.
9. Ball, R. C.; Richmond, P.; Phys. Chem. Liq. 1980, 9, 99.
10. Richardson, J. F.; Zaki, W. N. Chem. Eng. Sci., 1954, 3,
    65.
11. Carter, C.; Hibberd, D. J.; Howe, A. M.; Mackie, A.R.;
    Robins, M. M. Progr. Colloid Polym. Sci., 1988, 76,   37.
12. Gunning, P. A.; Hennock, M. S. R.; Howe, A. M.; Mackie, A. R.;
    Richmond, P.; Robins, M. M. Colloids and Surfaces , 1986, 20,
    65.
13. Gunning, P. A.; Hibberd, D. J.; Howe, A. M.; Robins,
    M. M. Food Hydrocolloids, 1988, 2, 119.
14. Michaels, A. S.; Bolger, J. C.; I&EC Fund. 1962, 1, 24.
15. Sperry, P. R.; Hopfenburg, H. B.; Thomas, N. L. J.Coll.
    Int. Sci., 1981, 82, 62.
16. Sperry, P. R. J. Coll. Int. Sci., 1982, 87, 375.

RECEIVED August 16, 1990

Chapter 18

Effects of Electrolyte on Stability of Concentrated Toluene in Water Miniemulsions

An Electroacoustic Study

Richard J. Goetz[1] and Mohamed S. El-Aasser[2]

[1]Department of Chemical Engineering, Emulsion Polymers Institute, and [2]Center for Polymer Science and Engineering, Lehigh University, Bethlehem, PA 18015

The electrokinetic sonic amplitude (ESA) is measured for toluene in water miniemulsions stabilized by cetyl alcohol (CA) and sodium lauryl sulfate (SLS). The general trend with increasing CA concentration is a sharp drop in the potential up to 20 mM CA, which remains relatively constant as the CA concentration is further increased. The general trends in the electrokinetic sonic amplitude (ESA) and ζ-potential calculated from the ESA agree with the ζ-potential determined with microelectrophoresis for o/w miniemulsions prepared at several CA concentrations. However, the magnitude of the microelectrophoretic ζ-potentials are approximately four times higher than the electroacoustic ζ-potentials. It is shown that the effects of electrolyte on the electrical properties of the miniemulsions can be followed through the effect on the ESA to the point of coagulation. The data clearly show maxima in the ESA as the electrolyte concentration is increased and is likely due to the displacement of CA adsorbed at the o/w interface by SLS.

Emulsion stability can be characterized through a variety of fundamental tests which attempt to analyze the durability of emulsions against phase separation. A few of these tests include ultracentrifugation, shelf life, and determining the critical coagulation concentration (CCC) with mono-, di-, and trivalent electrolytes. Important in emulsion stabilization is the magnitude of the charge separation that exists at the o/w interface, which originates from the adsorption of ionic surfactants. The magnitude of the surface charge represents a measure of the electrostatic repulsive forces, which is one of the primary stabilization mechanisms for most colloids. The electrical potential near the o/w interface of the droplet can be determined through microelectrophoresis experiments; however, the emulsion must be substantially diluted to approximately 100 ppm. Diluting an emulsion can change the chemical nature of the continuous phase, shift the surfactant adsorption equilibrium and ultimately affect the

0097–6156/91/0448–0247$06.00/0

stability of the emulsion. Thus, perturbing the chemical balance of the continuous phase upon dilution can result in erroneous electrophoretic potentials. This report is concerned with the measurement of electrokinetic potentials of concentrated o/w miniemulsions through a new technique, electroacoustics.

Electroacoustic measurements provide an unobtrusive method to examine the electrokinetic properties of concentrated dispersions. This technique has been recently used to study the effect of polymer adsorption on the ESA of aqueous silica dispersions (1), to determine the isoelectric point of alumina (2) to measure the point of zero charge of TiO_2 dispersions, and the mobility of bitumen emulsions at various pH's (3). However, this technique is ideally suited for observing changes in the electrokinetic potentials of concentrated o/w or w/o emulsions at their formulation concentration, since dilution is not required.

The colloid vibrational potential (CVP) and electrokinetic sonic amplitude (ESA) are the two potentials which are measured with electroacoustics. The CVP is obtained by applying an ultrasonic acoustic wave with a frequency in the MHz range to the dispersion. The difference in inertia between the colloid particle and ions in the double layer induces an alternating potential. Each particle in the dispersion acts as an alternating dipole moment, which sum to a larger potential. The CVP is defined as the amplitude of the electric field per unit velocity amplitude (of the applied ultrasonic field) with units of mV·s/m. This effect was first observed by Debye (4) for electrolyte solutions and was defined as the ion vibrational potential (IVP). Later, Rutgers (5) and Herman (6) showed that the effect was observable in colloidal dispersions, and Herman (6), Enderby (7), and Enderby and Booth (8) derived theoretical expressions for the CVP. Marlowe et al. (9) recently described a new apparatus that measures the CVP, using continuous wave ultrasonics, and derived an equation to relate the CVP to the ζ-potential based on the work of Enderby and Enderby et al. (7,8). Recently, O'Brien (10) derived expressions for both the CVP and ESA and demonstrated that the work of Enderby (7,8) was incorrect, since the authors didn't account for the complex conductivity and assumed that the electric field within the particle was unaffected by the applied field.

The ESA, which is the reciprocal effect of the CVP, was recently developed by Oja et al. (11). It is obtained by applying a high frequency alternating electric field, with a frequency on the order of 1 MHz, to the dispersion that forces the charged particles to undergo oscillatory motion. The dynamic motion of the particles in the dispersion generates an acoustic wave with the same frequency of the electric field (11). The ESA is defined as the pressure amplitude per unit electric field with units of mPa·m/V.

The equations developed by O'Brien (10) to obtain the dynamic mobility and ζ-potential from the ESA and CVP are limited to dilute dispersions, thin double layers ($\kappa a \gg 1$), and low Ψ, where Ψ is the surface potential. The dynamic mobility, $\mu(\omega)$, can be calculated from either the CVP or ESA (for parallel plate geometry):

$$\mu(\omega) = \frac{ESA(\omega)}{\phi \, \Delta\rho \, C} = \frac{CVP(\omega) \, K^*}{\phi \, \Delta\rho \, C} \qquad (1)$$

where $\mu(\omega)$ is the dynamic mobility in units of m^2/V s, ω is the frequency of the ultrasonic or electric field ϕ is the dispersion phase volume, $\Delta\rho$ is the density difference between the dispersed and continuous phase, C is the velocity of sound in the dispersion, and K^* is the complex conductivity of the dispersion. Because K^* cannot be determined with the instrument used (Matec Applied Sciences), only the ζ-potential determined from the ESA is compared.

The dynamic mobility, $\mu(\omega)$, has the same units as the low frequency mobility determined from microelectrophoresis and the two are similar when particle inertial effects are negligible. O'Brien's (10) relation between the ζ-potential and dynamic mobility for dilute dispersions is:

$$\zeta = \frac{\mu(\omega)\,\eta}{\varepsilon_0 D}\; \frac{1 - \frac{1}{9}i\alpha(3+\frac{\Delta\rho}{\rho})}{1+(1+i)\sqrt{\frac{\alpha}{2}}} \qquad (2)$$

where $\alpha = R^2\omega/v$, η is the solvent viscosity, ε_0 is the permitivity in a vacuum, D is the dielectric constant, v is the kinematic viscosity, and R is the radius of the colloid. The first term in Eq. [2] is Smoluchowski's equation for electrophoresis and the second term is a factor that accounts for the particle inertia and frequency dependence. Although electroacoustics is an ideal technique to measure electrokinetic potentials for concentrated dispersions, the theory is only valid at low solids content. Thus, the calculated values of $\mu(\omega)$ and ζ-potential from concentrated dispersions can only be compared on a relative basis at constant dispersion concentration.

Because the ESA is measured in concentrated emulsions, the effect of ϕ on the ultrasonic velocity, C, must be considered to calculate $\mu(\omega)$ from the ESA. Ultrasonic measurements are typically used to determined the phase volume and the adiabatic compressibility of colloidal dispersions and equations relating C to ϕ are available for a variety of systems (12). Urick (13) developed an equation relating ϕ to C for the case that the ultrasonic wavelength is greater than R (which is obeyed for this system). Urick's (13) relation is:

$$C(\phi) = \{((1-\phi)\rho_w + \phi\rho_0)((1-\phi)m_w + \phi m_0)\}^{-1} \qquad (3)$$

where m_w and m_0 and ρ_w and ρ_0 are the adiabatic compressibility and density of the aqueous and dispersed phases, respectively. Allinson (14) has shown that Eq. [3] accurately predicted $C(\phi)$ in a xylene in water emulsion up to a phase volume of 70%. Since the physical properties of xylene and toluene are similar, Eq. [3] is used to account for the effect of the dispersion phase volume of toluene miniemulsions on $C(\phi)$, using 9.245×10^{-10} m^2/N and 4.45×10^{-10} m^2/N for the adiabatic compressibility and 0.86 g/cm^3 and 1.0 g/cm^3 for the density of toluene and water, respectively.

The objective of this report is to examine the stability of concentrated toluene in water miniemulsions against electrolyte flocculation characterized through the electroacoustic potentials. Since this technique is relatively new, a preliminary investigation is conducted comparing the electroacoustic and electrophoretic potentials and the effects of dispersion concentration. Thus, a secondary objective of this work is to explore the use of electroacoustics to characterize the electrokinetic properties of o/w emulsions.

Materials and Methods

Toluene o/w miniemulsions were prepared from the dilute gel phase of cetyl alcohol (CA) and sodium lauryl sulfate (SLS), which has previously shown to form more stable emulsions than those prepared with SLS alone (15, 16). The SLS-CA gel phase was prepared by mixing the appropriate amount of water, CA, and SLS at 70°C for 2 hr and then cooling to room temperature. The miniemulsions were formed by emulsifying 180 ml of the gel phase and the appropriate amount of toluene with a Branson Heat Systems sonifier at power level 5 for a duration of 60 s at room temperature.

The electroacoustic measurements were preformed with Matec ESA-8000 system, which has been previously described by Babchin et al. (3). The sample volume varied between 180 to 260 ml. All measurements were made in a teflon beaker at 25°C fitted in the SSP-1 sample cell assembly. In the instrument used, the bulk conductivity is substituted for K^* to calculate $\mu(\omega)$ and the ζ-potential from the CVP. However, K^* can be experimentally measured through dielectric spectroscopy (17). The ESA mode of the instrument was calibrated daily using an aqueous 10% vol Ludox TM (Dupont). The ESA for this dispersion is -3.67 mPa m/V (18).

Electrophoretic mobilities of the miniemulsions were determined with the PenKem System 3000. The miniemulsions were first diluted to approximately 100 ppm with the serum of their continuous phase. Zeta potentials were then calculated from the mobility from the equations derived by O'Brien and White (19).

The droplet size distribution of the miniemulsion was obtained by dynamic light scattering with a Nicomp Model 370 submicron analyzer and transmission electron microscopy (TEM) of the stained droplets. For the light scattering measurements, the emulsion was diluted with an aqueous surfactant solution that matched the continuous phase of the emulsion. The droplet size distribution was determined with TEM by first staining the styrene in water miniemulsion with OsO_4. Styrene replaced toluene as the oil phase in the TEM work only, since toluene does not react with OsO_4. Approximately 0.4 ml of the emulsion was mixed with 0.2 ml of a 4.5% aqueous OsO_4 solution. The emulsion was then diluted and placed on a Formvar coated grid. The mean droplet radii was needed to calculate the ζ-potential from Eq. [2].

Results and Discussion

Figure 1 shows a plot of the ESA as a function of CA concentration for toluene in water miniemulsions prepared at eight toluene concentrations and 10 mM SLS. The ESA sharply decreased and approached a constant value at 20 mM CA, as the concentration of CA increased. This was observed at all toluene concentrations. The increase in the ESA with increasing toluene concentration at a given CA concentration is largely due the fact that the magnitude of the ESA is proportional to ϕ in addition to any changes in the surface charge that occurs with changes in ϕ. Similar trends are observed for the ζ-potential calculated from the ESA values in Figure 1, using Eqs. [1-3] as a function of CA concentration (Figure 2). However, the ζ-potential decreased as the toluene concentration increased. Because particle inertia effects are nearly negligible for this system, the trends in the ESA and electroacoustic ζ-potential should be similar. The effect of CA to partially reduce the electrophoretic ζ-potential has been previously shown by Elworthy et al. (20) for diluted chlorobenzene in water emulsions stabilized by a nonionic surfactant, $C_{16}EO_6$. However, the electrophoretic ζ-potentials determined from diluted xylene in water emulsions stabilized by CA and cetyl trimethyl ammonium bromide (CTAB) show the opposite behavior (21). The ζ-potential increased with increasing CA concentration when the total CTAB concentration was above its CMC. At CTAB concentrations below the CMC, a systematic trend was not observed. The above results suggest that the effect of CA on the adsorption at the o/w interface is dependent on the sign of the hydrophilic moiety of the surfactant.

Figure 3 compares the ζ-potentials determined from microelectrophoresis and electroacoustic as a function of CA concentration for miniemulsions prepared at 25% vol. toluene and 10 mM SLS. The microelectrophoresis ζ-potentials were from miniemulsions prepared at 25% vol then diluted to 100 ppm with serum from their continuous phase. The results show that the potentials determined from microelectrophoresis are approximately four times higher than the electroacoustic ζ-potentials. This difference is largely due to the effects of dispersion concentration, which are not taken into account in O'Brien's (10) derivation, Eqs. [1-2]. Zukoski et al. (22, 23) has recently shown that particle-particle interactions cancel for concentrated dispersions undergoing electrophoretic motion and the decrease in the mobility is solely due to the backflow of fluid to conserve volume. This leads to a linear relationship between the electrophoretic mobility and ϕ. Although the magnitude of the potentials are not similar, it is notable that the trends with increasing CA concentration are the same. Currently, experiments are being conducted to examine the effect of toluene concentration on the electroacoustic potentials of o/w miniemulsions.

Additionally, any difference in the position of the shear plane sensed by electroacoustics and electrophoresis would contribute to the difference in the ζ-potentials. The above results suggest that the electrophoretic shear plane is closer to the particle's surface. Assuming the electrophoretic shear plane is 2 Å away from the droplet surface, the electroacoustic shear plane would be located a few nanometers from the surface. This distance is unreasonably high. However, this calculation excludes the contribution of concentration effects to the difference in the ζ-potentials, which need to be considered.

Figure 4 shows the ESA as a function of NaCl concentration for miniemulsions prepared with 10 mM SLS and 30 mM CA at six different toluene phase volumes. Figure 5 shows the corresponding ζ-potentials calculated from the ESA values from three of the emulsions in Figure 4, using Eqs. [1-3]. Surprisingly, the data show an increase in the surface charge of the emulsion droplets as the concentration of NaCl increases, and reaches a maximum near 10 mM NaCl. Beyond the maxima, the surface potential then decreases with the collapse of the double layer to

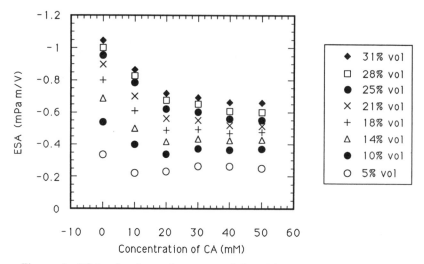

Figure 1. ESA of toluene in water miniemulsions prepared with 10 mM SLS as a function of CA concentration at eight different toluene phase volumes.

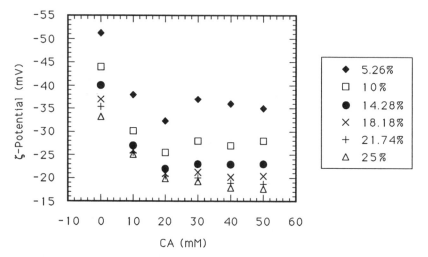

Figure 2. ζ-potential of toluene in water miniemulsions prepared with 10 mM SLS as a function of CA concentration at six different toluene phase volumes calculated from the ESA in Figure 1, using Eqs. [1-3].

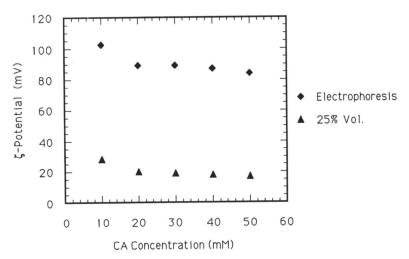

Figure 3. A comparison of the ζ-potential determined from microelectrophoresis and electroacoustics at 25% vol toluene as a function of CA concentration for miniemulsions prepared with 10 mM SLS 25% vol toluene.

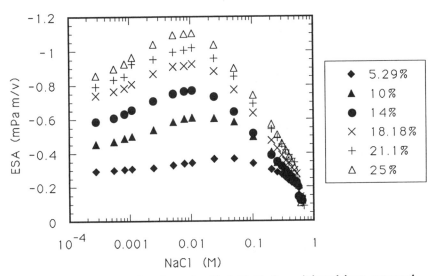

Figure 4. ESA as a function of added NaCl for miniemulsions prepared with 10 mM SLS and 30 mM CA at six different toluene concentrations.

the point of coagulation. Increases in the ζ-potentials of emulsions droplets, prior to the point of charge reversal, have been observed in other emulsion systems; however, an explanation has never been given (24, 25). Similar behavior has also been observed in polystyrene latexes and several explanations for the maxima in these systems exist.

The adsorption of co-ions has been suggested as the source of surface charge on non-ionogenic surfaces (26) and has been used to explain the origin of the ζ-potential/electrolyte maxima (27, 28). Maxima have been observed on polystyrene latex with surface sulfate groups and latex in which the sulfate groups were completely hydrolyzed (27). A different perspective to the co-ion adsorption mechanism was recently suggested by Miklavic et al. (29). As the concentration of ions is increased, the H^+ adsorbed in the inner Helmoltz plane was replaced by a larger positive ion. The replacement of H^+ ions by the larger ion re-exposes charged surface sites and results in an increase in the electrokinetic potential. Thus, the increase in surface charge with increasing electrolyte concentration is due to the inability of the larger counter-ion to neutralize the surface charge.

A counter argument to the co-ion adsorption model is sometimes referred to as the hairy layer or surface roughness model (28, 30-31). This model considers the surface to be covered with polyelectrolyte chains carrying the surface charge. This layer contracts with increasing electrolyte concentration, shifting the electrokinetic shear plane closer to the particle surface. This would then increase the electrokinetic potential. Midmore et al. (31) have suggested that the dominant contribution to the maxima is the introduction of ionic

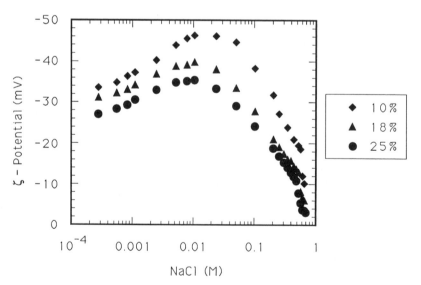

Figure 5. ζ-potential as a function of added NaCl for miniemulsions prepared with 10 mM SLS and 30 mM CA at three different toluene concentrations calculated from the ESA values in Figure 4., using Eqs. [1-3].

conduction in the diffuse layer inside the shear plane at low electrolyte concentration. However, this mechanism can be ruled out for the miniemulsions used in this study. The low HLB (hydrophile-lipophile balance) and molecular weight of CA would prevent an adsorption configuration which would promote a surface roughness or hairiness on the miniemulsion droplet.

Figure 6 shows a comparison of the ESA as a function of NaCl for a miniemulsion prepared with 10 mM SLS and 50 mM CA and an emulsion prepared with only 10 mM SLS each at 25% vol. toluene. The data show that the ESA/electrolyte maxima is only observable when CA is present. This suggests that the mechanism for the maxima is either due to the displacement of CA adsorbed at the o/w interface from SLS in the continuous phase or co-ion adsorption. The addition of electrolyte decreases the solubility of SLS in the continuous phase, favoring adsorption. Since SLS is a stronger amphiphile than CA, the adsorbed CA is displaced by SLS. Additionally, maxima have been observed with polystyrene latex in which the surface had been completely hydrolyzed, thus, co-ion adsorption cannot be ruled out.

The displacement mechanism would also explain the absence of an ESA/electrolyte maxima when CA is not present in the emulsion. For emulsions prepared without CA, the interfacial area would need to increase for the surface excess of SLS to increase, which is energetically unfavorable. It is likely that the addition of electrolyte could reduce the adsorption area of SLS at droplet surface, providing surface for additional surfactant from the continuous phase. However, electrolyte tends to dehydrate the surface which would reduce the surface charge.

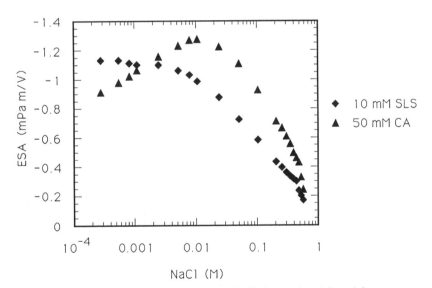

Figure 6. ESA as a function of added NaCl for a a) miniemulsion prepared with 10 mM SLS and 50 mM CA and b) only 10 mM SLS, each at 25% vol toluene.

Figure 7 shows an expanded region of Figure 5 focussed at high electrolyte concentration near the point of coagulation. The collapse of the double layer with increasing electrolyte concentration results in an exponential decrease in the ESA. As the electrolyte concentration increases, the double layer collapses to the point where the Van der Waals attractive forces predominate over the electrostatic repulsive forces. At this point the emulsion coagulates, resulting in a sharp drop in the ESA. The electrolyte concentration at which this occurs is defined as the critical coagulation concentration (CCC).

Figure 8 shows the CCC as a function of CA concentration for miniemulsions prepared with 10 mM SLS at 25% vol toluene. The results indicate that miniemulsion stability against electrolyte coagulation increases with increasing CA concentration. This likely originates from the change in the interfacial composition, i.e., surface charge, activated by the initial addition of electrolyte. Figure 9 shows the CCC of miniemulsions prepared with 10 mM SLS and 30 mM CA as a function of toluene concentration. Figure 9 shows that the miniemulsion stability decreases with increasing toluene phase volume. As the electrostatic repulsive forces decrease with increasing NaCl concentration, the emulsion enters the regime of rapid coagulation, i.e., each collision results in the formation of a coalesced droplet. The rate of rapid coagulation is proportional to ϕ^2, thus, the point of critical coagulation is likely to occur at a lower electrolyte concentration. However, the exact mechanism for the stability with increased CA and toluene concentration is difficult to pinpoint, since the physical properties, i.e., size, number density, and surface charge, of the emulsion are dependent on these factors. These properties are coupled together in defining the overall stability of the miniemulsion.

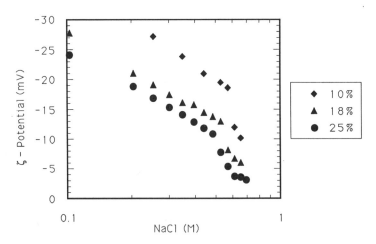

Figure 7. An enlarged region of Figure 5 near the critical coagulation concentration.

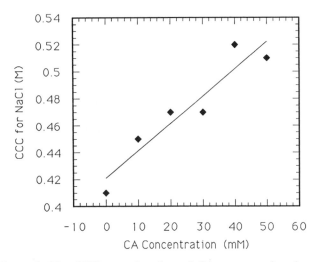

Figure 8. The CCC as a function of CA concentration for miniemulsions prepared with 10 mM SLS and 25% vol. toluene.

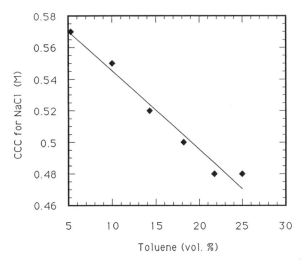

Figure 9. The CCC as a function of toluene concentration for miniemulsions prepared with 10 mM SLS and 30 mM CA.

Summary

Electroacoustic measurements provide an unobtrusive means to measure the relative change in the surface charge properties of concentrated dispersions. This is ideally suited for miniemulsions, since dilution is not required, allowing them to be examined at their formulation concentration. However, only the relative changes at constant dispersion concentration can be compared. This is due to the effects of dispersion concentration, which is illustrated when comparing the electroacoustic and electrophoretic potentials.

The effect of electrolyte on the ESA shows a clear increase in the ESA at an NaCl concentration in the range of 10^{-2} M NaCl. It is suggested that this is due to the displacement of adsorbed CA by SLS from the continuous phase. The ESA then decreases with the collapse of the electrical double layer, as the NaCl concentration is further increased.

The CCC values for NaCl indicate that emulsions prepared at the highest CA concentration up to 50 mM and the lowest possible phase volume have the highest stability against electrolyte coagulation.

Legend of Symbols

English
C, C(ϕ)   ultrasonic velocity (m/s)
CVP   Colloid Vibrational Potential (mV m/s)
D   dielectric Constant
ESA   Electrokinetic Sonic Amplitude (mPa m/V)
K*   Complex Conductivity
m   adiabatic compressibility (m^2/N)
R   droplet radius (m)

Greek
α $R^2\omega/\nu$
$\Delta\rho$ $\rho_w-\rho_o$ (Kg/m^3)
ε_0 permitivity in a vacuum (C^2/J m)
ζ zeta potential (V)
η solvent viscosity (kg/m s)
$\mu(\omega)$ dynamic mobility (m^2/Vs)
ν kinematic viscosity (m^2/s)
ρ solvent density (kg/m^3)
ϕ dispersion volume fraction
ω frequency (Hz)

Subscripts

o   dispersed phase (oil)
w   continuous phase (water)

Acknowledgments

Support from the Colgate-Palmolive Co. and the Plastics Institute of America, Inc. and the use of the Matec ESA 8000 from Matec Applied Science is gratefully appreciated.

Literature Cited

1. Maier, H.; Baker, J.; Berg, J.C. J. Coll. Interface Sci. 1983, 119, 312.
2. ESA Application Note, CMS 350, Matec Applied Sciences, Hopkinton, MA,1985.
3. Babchin, A.J.; Chow, R.S.; Sawatzky, R.P. Adv. Coll. Interface Sci. 1989, 30, 111.
4. Debye, P., J. Chem. Phys. 1933, 1, 13.
5. Rutgers, A. Physica 1938, 5, 674.
6. Hermans, J. Phil. Mag. 1938, 25, 426.
7. Enderby, J.A. Proc. Roy. Soc. 1957, A207, 329.
8. Booth, C. ; Enderby, J. Proc. Phys. Soc. 1957, A208, 357.
9. Marlowe, B.J. ; Fairhurst, D. Langmuir 1988, 4, 611.
10. O'Brien, R.W. J. Fluid Mech. 1988, 190, 71.
11. Oja, T.; Petersen, G.L.; and Cannon, D.W. US Patent 4 497 207, 1985.
12. McClement, P.J. Adv. Coll. Interface Sci. 1987, 27, 285.
13. Urick, J. J. Applied Phys. 1947, 18, 933.
14. Allinson, P.A. J. Coll. Interface Sci. 1958, 13, 513.
15. Lack, C. Ph.D. Dissertation, Lehigh University, Bethlehem, PA,1985.
16. El-Aasser, M.S.; Lack, C.D.; Choi, Y.T.; Min, T.I.; Vanderhoff, J.W. ; Fowkes, F.M. Colloid Surf. 1984, 12, 79.
17. Meyers, D.F. ; Saville, D.A. J. Coll. Interface Sci. 1989, 131, 448, 461.
18. O'Brien, R.W. ; White, L.R. J. Chem. Soc. Faraday Trans. 2 1978, 74, 1607.
19. Mann, R. personal correspondence, 1990.
20. Elworthy, P.H.; Florence, A.T. ; Rogers, J.A. J. Coll. Interface Sci. 1971, 35, 34.
21. Tadros, Th. F. Colloid Surf. 1985, 1, 83.
22. Zukoski, C.F. ; Saville, D.A. J. Coll. Interface Sci. 1987, 115, 422.
23. Zukoski, C.F. ; Saville, D.A. J. Coll. Interface Sci. 1989, 132, 222.
24. Usui, S., ; Imamura, Y. J. Disper. Sci. Tech. 1987, 8, 359.
25. Prakash, C. ; Srivastava, N. Bull. Chem. Soc. Japan 1967, 40, 1756.
26. Abramson, H.A. In Electrokinetic Phenomena ACS Monograph Ser. No. 66; The Chemical Catalog Co., New York 1934; p 331.
27. Ma, C.M.; Micale, F.J.; El-Aasser, M.S.; Vanderhoff, J.W. In Emulsion Polymers in Emulsion Polymerization, Basset, D.R.; Hamielic, A.E., Eds.; ACS Symposium Series No. 165; American Chemical Society: Washington D.C. 1981; p 252.
28. Elimelech, M.; O'Melia, C.R. Colloid Surf. 1990, 44, 165.
29. Millavic, S.J. ; Ninham, B.W. J. Coll. Interface Sci. 1990, 134, 325.
30. Van der Put, A.G.; Bijsterbosch, B.H. J. Coll. Interface Sci. 1983, 92, 499.
31. Midmore, B.R. ; Hunter, R.J. J. Coll. Interface Sci. 1988, 122, 521.

RECEIVED August 7, 1990

Author Index

Affiliation Index

Subject Index

A

2-Acyl glycerols
preparation problems, 52
preparation via triglyceride hydrolysis, 52f,53
regioselective esterification, 51,52f
Amphiphilic lipopeptides, formation, 103
1-Anilinonaphthalene-8-sulfonic acid, use
as extrinsic fluorescent probes,
200 – 201

Aqueous dispersions, types of lipid phases
used in formation, 44
Arlacel 20, effectiveness as cosurfactant
in microemulsions, 74
α-tending emulsifier(s)
aggregation of fat globules, 146 – 147
characterization, 147
examples, 146 – 147
formation of α-gel phase, 147

Production: Victoria L. Contie
Indexing: Deborah H. Steiner
Acquisition: Barbara C. Tansill

Books printed and bound by Maple Press, York, PA

*Paper meets minimum requirements of American National Standard
for Information Sciences—Permanence of Paper for Printed Library
Materials, ANSI Z39.48–1984* ∞

Other ACS Books

Chemical Structure Software for Personal Computers
Edited by Daniel E. Meyer, Wendy A. Warr, and Richard A. Love
ACS Professional Reference Book; 107 pp;
clothbound, ISBN 0–8412–1538–3; paperback, ISBN 0–8412–1539–1

Personal Computers for Scientists: A Byte at a Time
By Glenn I. Ouchi
276 pp; clothbound, ISBN 0–8412–1000–4; paperback, ISBN 0–8412–1001–2

Biotechnology and Materials Science: Chemistry for the Future
Edited by Mary L. Good
160 pp; clothbound, ISBN 0–8412–1472–7; paperback, ISBN 0–8412–1473–5

Polymeric Materials: Chemistry for the Future
By Joseph Alper and Gordon L. Nelson
110 pp; clothbound, ISBN 0–8412–1622–3; paperback, ISBN 0–8412–1613–4

The Language of Biotechnology: A Dictionary of Terms
By John M. Walker and Michael Cox
ACS Professional Reference Book; 256 pp;
clothbound, ISBN 0–8412–1489–1; paperback, ISBN 0–8412–1490–5

Cancer: The Outlaw Cell, Second Edition
Edited by Richard E. LaFond
274 pp; clothbound, ISBN 0–8412–1419–0; paperback, ISBN 0–8412–1420–4

Practical Statistics for the Physical Sciences
By Larry L. Havlicek
ACS Professional Reference Book; 198 pp; clothbound; ISBN 0–8412–1453–0

The Basics of Technical Communicating
By B. Edward Cain
ACS Professional Reference Book; 198 pp;
clothbound, ISBN 0–8412–1451–4; paperback, ISBN 0–8412–1452–2

The ACS Style Guide: A Manual for Authors and Editors
Edited by Janet S. Dodd
264 pp; clothbound, ISBN 0–8412–0917–0; paperback, ISBN 0–8412–0943–X

Chemistry and Crime: From Sherlock Holmes to Today's Courtroom
Edited by Samuel M. Gerber
135 pp; clothbound, ISBN 0–8412–0784–4; paperback, ISBN 0–8412–0785–2

For further information and a free catalog of ACS books, contact:
American Chemical Society
Distribution Office, Department 225
1155 16th Street, NW, Washington, DC 20036
Telephone 800–227–5558

Im